Solar Power for Your Home

Other McGraw-Hill Books of Interest

Do-It-Yourself Home Energy Audits: 140 Simple Solutions to Lower Energy Costs, Increase Your Home's Efficiency, and Save the Environment by David S. Findley

Renewable Energies for Your Home: Real-World Solutions for Green Conversions by Russel Gehrke

Build Your Own Plug-In Hybrid Electric Vehicle by Seth Leitman

Build Your Own Electric Bicycle by Matthew Slinn

Build Your Own Electric Motorcycle by Carl Vogel

Solar Power for Your Home

David S. Findley

New York Chicago San Francisco
Lisbon London Madrid Mexico City
Milan New Delhi San Juan
Seoul Singapore Sydney Toronto

The McGraw-Hill Companies

Cataloging-in-Publication Data is on file with the Library of Congress

McGraw-Hill books are available at special quantity discounts to use as premiums and sales promotions, or for use in corporate training programs. To contact a representative, please e-mail us at bulksales@mcgraw-hill.com.

Solar Power for Your Home

1 2 3 4 5 6 7 8 9 0 DOC/DOC 1 6 5 4 3 2 1 0

ISBN 978-0-07-166784-5
MHID 0-07-166784-9

The pages within this book were printed on acid-free paper containing 100% postconsumer fiber.

Sponsoring Editor
Judy Bass

Editorial Supervisor
Stephen M. Smith

Production Supervisor
Pamela A. Pelton

Acquisitions Coordinator
Michael Mulcahy

Project Manager
Patricia Wallenburg, TypeWriting

Copy Editor
Lisa Theobald

Proofreader
Paul Tyler

Indexer
Jack Lewis

Art Director, Cover
Jeff Weeks

Composition
TypeWriting

To you, love always and all ways!

I would like to thank my brother Gordon for helping in my time of need,
and assuming responsibility after my demise.

"In the end, it's not the years in your life that count.
It's the life in your years."

Abraham Lincoln

I would like to thank the wonderful people without whom
my two books would not have been possible:

I am grateful to Carl Vogel for the introduction
and Seth Leitman for the opportunity.

I am especially grateful to Patty Wallenburg of TypeWriting
and Judy Bass of McGraw-Hill, who transformed my thoughts
into documents and my documents into books, and for
making dreams into opportunities and opportunities into reality.

"Books serve to show a man that those original thoughts
of his aren't very new at all."

Abraham Lincoln

About the Author

David S. Findley is a former professor at Farmingdale State College, Farmingdale, New York, and the owner of Synergy New Technology, a not-for-profit green technology solution provider. He has also overseen the development of a green engineering sciences curriculum. This is his second McGraw-Hill book, the first being *Do-It-Yourself Home Energy Audits*.

Contents

CHAPTER 1
The History of Solar Energy

If you believe the history of solar energy begins in the 1970s, around the time President Jimmy Carter installed solar panels in the White House (see Figure 1-1), you have underestimated the sun's history by a few billion years. In fact, life on Earth owes a great debt, if not its total existence, to solar energy. The sun is responsible for all life on our planet, including us humans.

Fact is, the sun provides more energy in one hour than all of humanity uses, in all forms, in a single year.

Sunlight and Life on Earth

Life is believed to have existed as early as 3.5 billion years ago. A single-celled, blue-green cyanobacteria, shown in Figures 1-2 and 1-3, flourished in the sunlit parts of the oceans.

Trillions of these microscopic organisms have transformed our planet. They capture and use the energy from the sun to create food, and they release oxygen as a waste product. For millions of years, cyanobacteria has changed the Earth's atmosphere from CO_2 to oxygen.

Scientists believe that around 3 billion years ago, autotrophic animals (such as bacteria) diversified from earlier species. These autotrophs were capable of synthesizing energy from complex inorganic material—that is, via the sun, photosynthesis, and other inorganic elements. These living organisms were able to tap into a completely new energy resource that was virtually inexhaustible: the sun.

Autotrophs, like cyanobacteria, produced substances required for human life. These bacteria fed on hydrogen sulfide, ammonium, and iron, and they produced oxygen.

FIGURE 1-1 President Jimmy Carter and the installation of solar panels on the White House
http://farm3.static.flickr.com/2674/3847362668_6ef6cff8d4.jpg

During the course of millions of years, autotrophs created an environment that allowed the evolution of life as we know it today. Sunlight allowed these organisms to convert inorganic materials into useful resources for life. Without the sun, it is almost certain that no life would have developed on our planet.

Autotrophs ruled the world until they created so much organic waste that they polluted their own environment, so that only limited amounts of autotrophs could survive. The extinction of the autotrophs led to one of the Earth's many "mass extinctions." The organisms died en masse, creating organic material that would be stored near the Earth's surface.

During the last 500 million years or so, five mass extinctions have occurred on our planet, as shown in Figure 1-4. The last is perhaps the best known to most of us—extinction of the dinosaurs. Each mass extinction killed much of the organic life on the Earth, including animals, plants, and bacteria. The organic material was buried and became the stored solar energy we know and use today—in the form of oil, coal, and natural gas. All the fossil fuels we know, and abuse, today were once organic material that used sunlight to become inorganic. Without the sun, we would have no supplies of fossil fuels.

White House Solar Panels

President Carter added solar panels on the roof of the White House, but President Ronald Reagan removed those same panels later. The panels did not go to waste, however, and were moved to Unity College in Unity, Maine, as shown here.

White House solar panels installed at Unity College
http://www.unity.edu/uploadedImages/wwwunityedu/EnvResources/Sustainability/image004.jpg

FIGURE 1-2 Oxygen-producing cyanobacteria
http://gallery.usgs.gov/images/12_07_2009/s85Are1QPk_12_07_2009/medium/Microcystis_in_Sytox_Green–Barry_Rosen.jpg

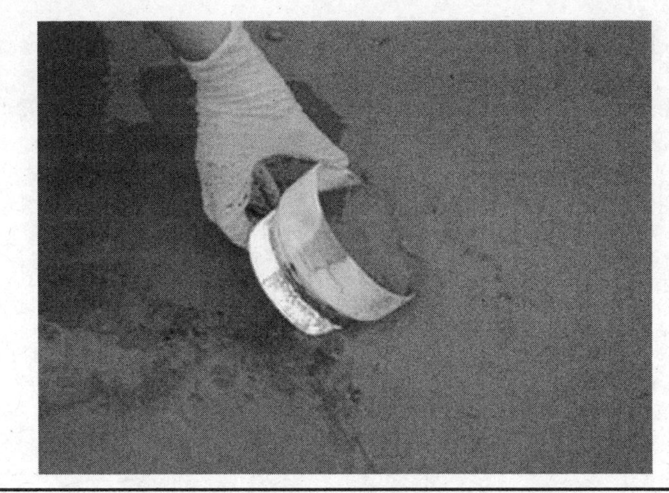

FIGURE 1-3 Waterborne cyanobacteria
http://ks.water.usgs.gov/studies/qw/cyanobacteria/binder-lake-ia.jpg

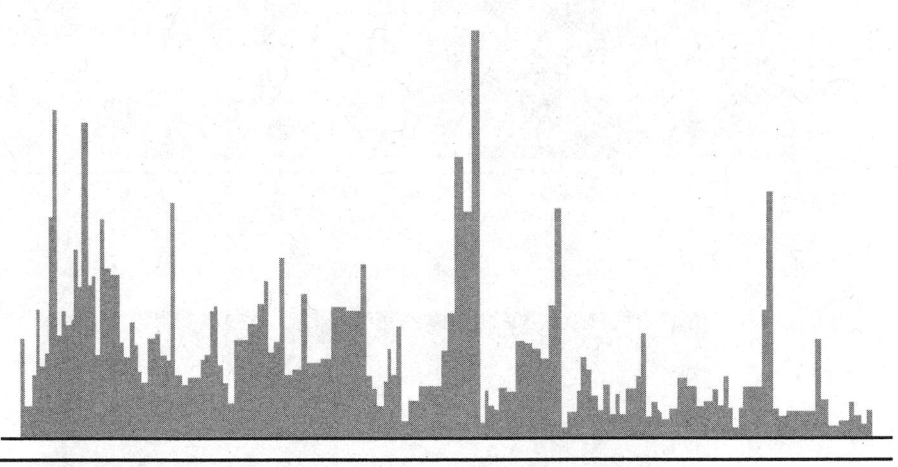

FIGURE 1-4 Five mass extinction events
http://en.wikipedia.org/wiki/File:Extinction_intensity.svg

The Human Factor

Humans have been spectators for only a brief instant in the world's history. *Homo sapiens* have been walking the Earth for approximately 200,000 years. For most of this time, humans were hunter-gatherers, surviving by hunting animals and gathering and eating foods that were nourished by none other than sunlight. About 10,000 years ago, humans began to employ agriculture to grow their own food. During this period, human food provided by hunting and gathering was supplemented with crops grown by early humans.

By around 10,000 B.C., the world's total population is believed to have grown to 4 to 6 million people. Figure 1-5 shows one estimate of the growth of human population through history.

Fast-forward to 14th-century Europe, when overcrowding and poor living conditions spread the Black Death, or bubonic plague, which reduced the world's population by about 100 million people. By 1500, the world's population had reached almost 500 million. And by the 1800s, the world's population had reached approximately 1 billion.

Then, as now, people were supported by sunlight. Solar energy was used as heat and light. Most jobs and tasks were performed during daylight hours. Early to bed, early to rise was a mandate, prescribed by the natural light. The food grown was supported by simple animal fertilizers, watered by the rain and crude irrigation, and nourished by sunlight (Figure 1-6). Limited areas on the planet could support a narrow quantity and type of crops, subject to minor climate variations.

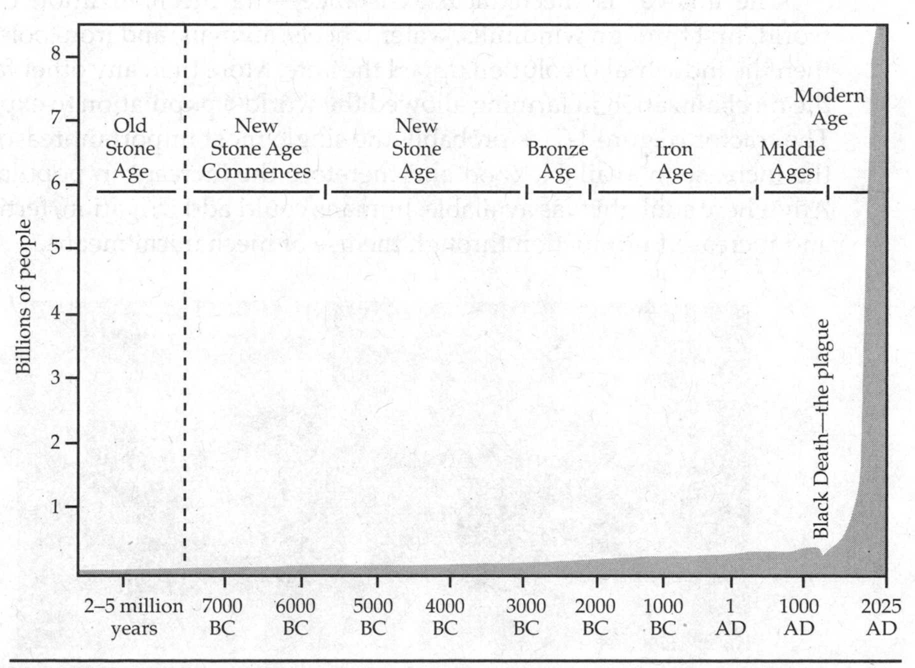

FIGURE 1-5 World population graph
http://www.susps.org/images/worldpopgr.gif

FIGURE 1-6 Sunlight in a forest
http://earthobservatory.nasa.gov/Experiments/ICE/panama/Images/panama_canopy.jpg

By 1950, the population had grown to more than 2.5 billion people. If you think about this fact in terms of the human timeline, you'll see that the population grew in 200,000 years to 1 billion people, and then, in only 150 years, another billion people were added to the planet. How did this occur? And why did this occur after hundreds of thousands of years of stable population growth?

The answer is mechanical assistance—the mechanization of the world, first through windmills, water wheels, animals, and iron tools, and then the industrial revolution upped the ante. More than any other factor, the mechanization in farming allowed the world's population to explode. The tractor (Figure 1-7) is probably the single most important reason for the increase in available food and therefore the increase in population. Anywhere sunlight was available, humans could add irrigation, fertilizer, and increased production through the use of mechanical means.

FIGURE 1-7 Tractors are the most important factor in increasing food yields.
http://www.fsa.usda.gov/Internet/FSA_MediaGallery/tractor_ut332933191.jpg

Figure 1-8 The 1970s gas shortage
http://www.nrel.gov/data/pix/Jpegs/00573.jpg

From 1950 to 1970, the population grew by another billion people. From 1970 to 1985, another billion people contributed to the world population, then at almost 5 billion. All competed for the same finite resources from a limited planet, a fact that became all too apparent during the 1970s, when the United States and other countries experienced gasoline shortages (Figure 1-8).

The Future

Events in our past can help foretell our future. Today, the world, and especially the United States, depends heavily on the fuel products created by sunlight—coal, gas, and oil. What if tomorrow you awoke to find no water in your home? No heat or electricity in your house? No food being delivered to the supermarket because no fuel was available for the trucks that deliver it? This is our imminent future if we do not change how we live. Producing the oil and coal we use today took the Earth and sun hundreds of millions of years and five mass extinctions (Figure 1-9).

The amount of oil that remains underground and available is debatable (Figure 1-10). But most scientists believe that fossil fuels, and that the Earth's ability to create them, is limited. That said, it seems reasonable to depend on the sun as a source of more energy. We should be using more sunlight—not fossil fuels—as a nonpolluting, ever available, energy form.

By 2011, the world's population is estimated to reach 7 billion people. We have been overusing our stored energy reserves to the point at which we may lose them. We are not able to grow sufficient amounts of food to support the current world population, which is expected to grow to more than 9 billion people by 2050 (Figure 1-11).

All of this points to one simple fact: We have been completely dependent on sunlight, stored or fresh, for all of humanity and all of life. And the sun is one resource we can always count on.

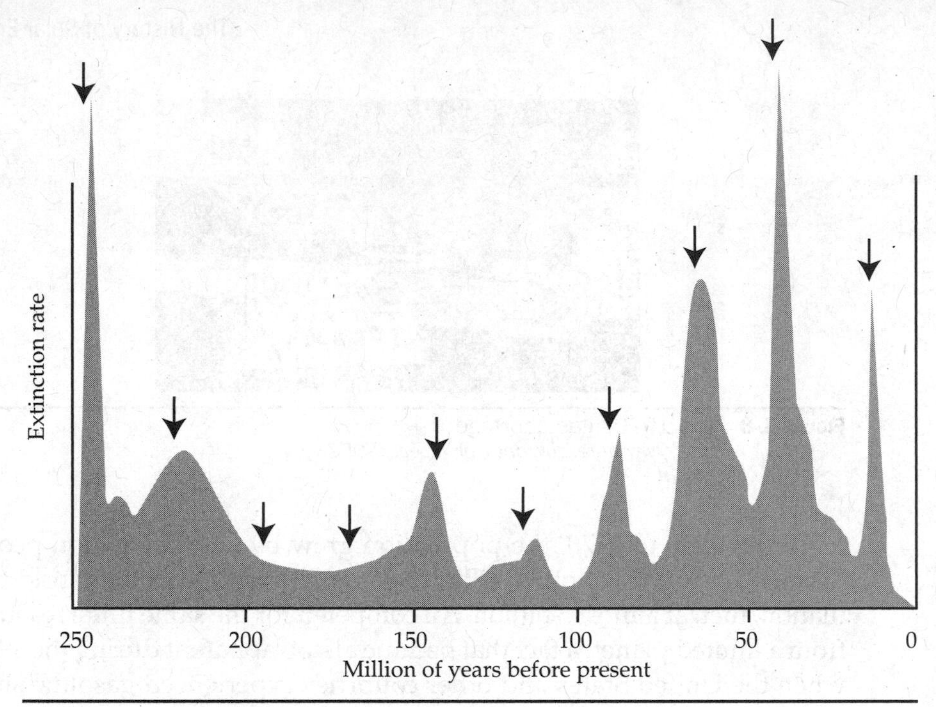

FIGURE 1-9 Five major mass extinctions
http://www.lbl.gov/Science-Articles/Archive/images5/raup-sepkoski-plot.gif

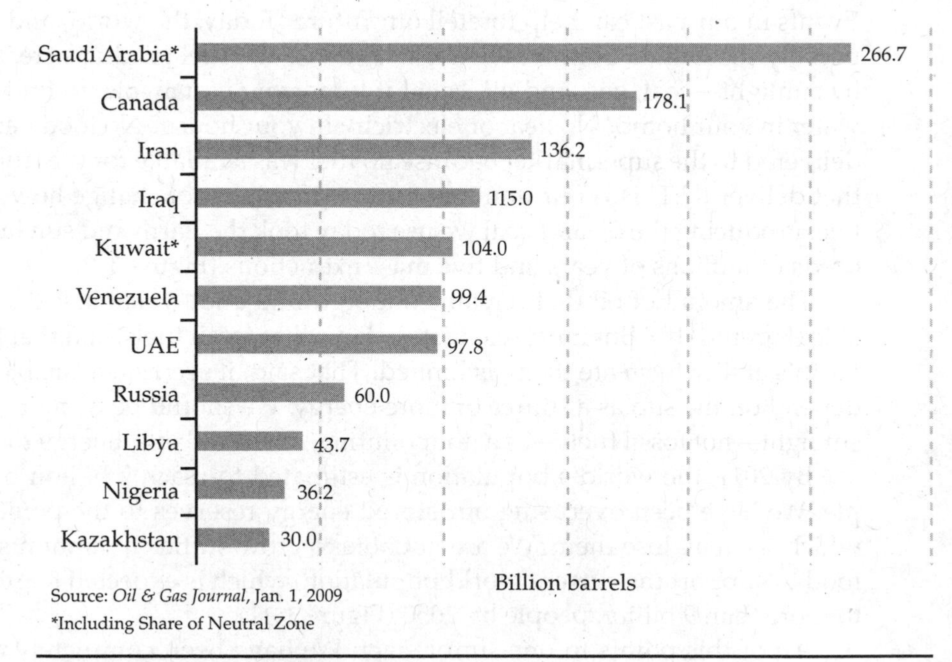

FIGURE 1-10 The world oil reserves
http://www.eia.doe.gov/emeu/cabs/Kuwait/images/09-03-kuwait-worldreserves.gif

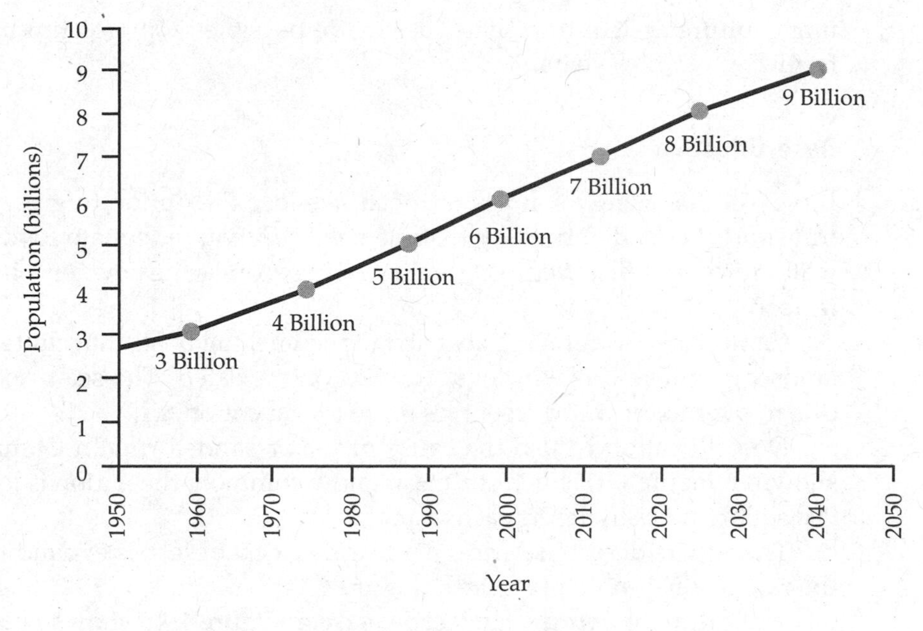

Source: U.S. Census Bureau, International Database, December 2008 update.

FIGURE 1-11 Projected world population
http://www.census.gov/ipc/www/img/worldpop.gif

Food Production and Declining Natural Resources

The chemicals, herbicides, bioengineered seed, fuel, and mechanization used in farming today are completely dependent upon an energy-rich society. Industrialized agriculture demands fossil fuels in two essential ways: in their direct consumption on the farm, and indirectly in their consumption by manufacturing. Direct consumables include fuels, lubricants, farm vehicles, and machinery. Farms use oil, liquid propane, natural gas, and electricity to power dryers, pumps, lights, heaters, coolers, and other machinery.

Indirect consumables comprise mainly oil, natural gas, and coal used to manufacture fertilizers, pesticides, and packaging and allow transportation, consumption, and waste disposal of food and non-food items. Because of our dependence on fossil fuels, an oil shortage would definitely impact our food supply.

Some farmers using modern organic farming methods have reported yields as high as those available from conventional farming—without the use of synthetic fertilizers and pesticides. The organic crops also have shown to be more resilient, drought-resistant, and nutrient-rich. However, the reconditioning of soil to restore nutrients lost during the use of monoculture agriculture techniques made possible by petroleum-based fertilizers requires years.

Desertification

If we lose the energy sources we need, we lose the option to grow and transport the food and water that we need. Without adequate food and water resources, *desertification* (Figure 1-12) becomes the new term in our lexicon.

Camels are now grazing near abandoned fishing boats that clutter the landscape where Kazakhstan's Aral Sea once existed. The sea dried up due to overuse of water resources for growing cotton.

Desertification is also the cause of major sandstorms in China, as shown in Figure 1-13. Such storms are now common; the nation is losing 900 square miles to desert each year.

The Sahara desert in Africa, the world largest desert, is expanding at the rate of a half-mile per month (Figure 1-14).

The future of Africa's landscape is clear. Figure 1-15 shows areas of Africa predicted to suffer from desertification. (Africa is the world's poorest continent with the highest birth rate.)

The United States is not immune to climate change, drought, or food shortages. Severe drought and desertification across the nation have permanently changed agriculture and grazing lands (Figure 1-16).

And with desertification comes mass migration, a phenomenon that began more than 20 years ago in Africa, as people have begun migrating from the areas of desertification to areas of opportunity. Often, as in Africa, drought and desertification first lead to famine and then lead to

FIGURE 1-12 Desert ships
http://ocho.uwaterloo.ca/~pfieguth/Personal/EnergyLimits/Figures/desert-ships.gif

FIGURE 1-13 Dust storm in Beijing, China
http://earthobservatory.nasa.gov/images/imagerecords/14000/14861/China_TMO_2005118_lrg.jpg

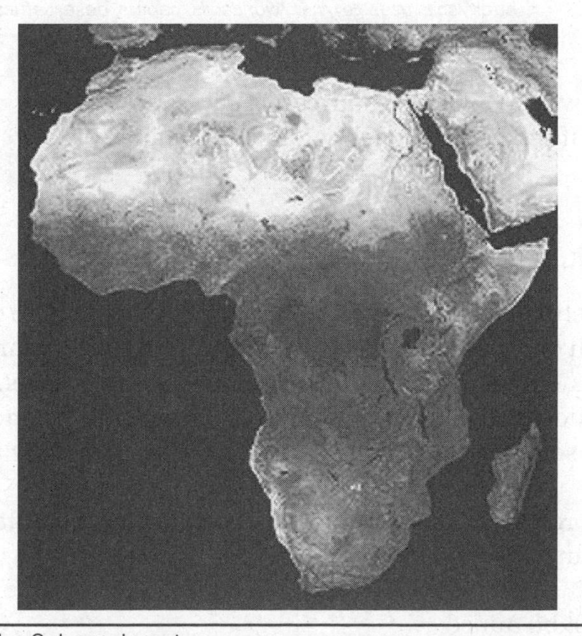

FIGURE 1-14 The Sahara desert
http://rst.gsfc.nasa.gov/Sect6/Africa_satellite_plane-225.jpg

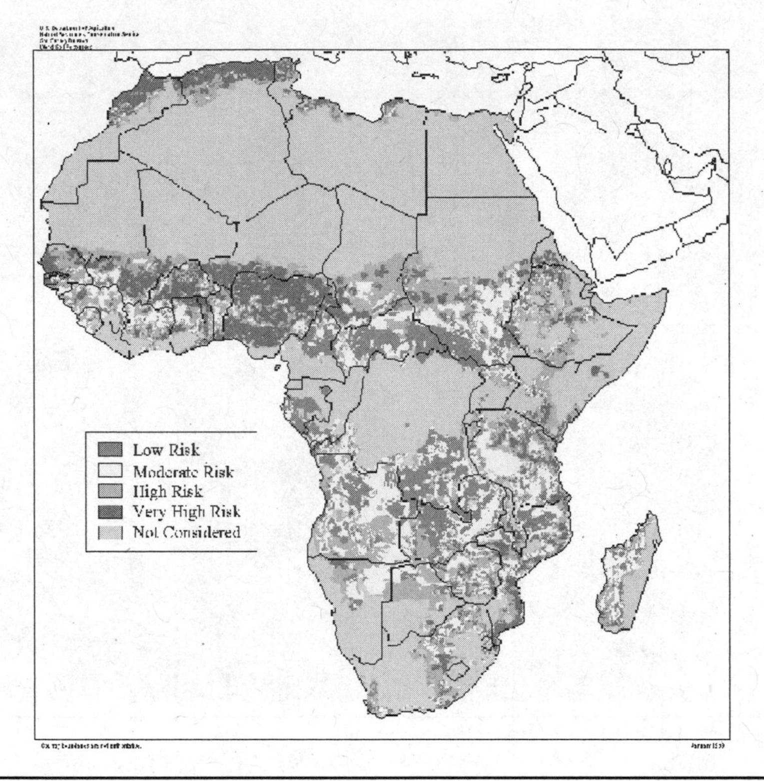

FIGURE 1-15 Desertification of Africa
http://soils.usda.gov/use/worldsoils/papers/desert-africa-fig2.gif

wars over available resources. Africa shows what can occur in other places if the land is no longer able to sustain life.

Solar for Humanity

Let's return to this statement: The sun provides more energy in one hour than all of humanity uses, in all forms, in a single year. Sunlight can provide us with its own resolution to our energy problems. The only transformation required is for humanity to reduce, or end, consumption of stored solar (as fossil fuels) and, in its place, use freely available "fresh" solar.

Stored solar, or fossil fuels, have many disadvantages, including the following:

- Must be mined
- Destroys and pollutes the land

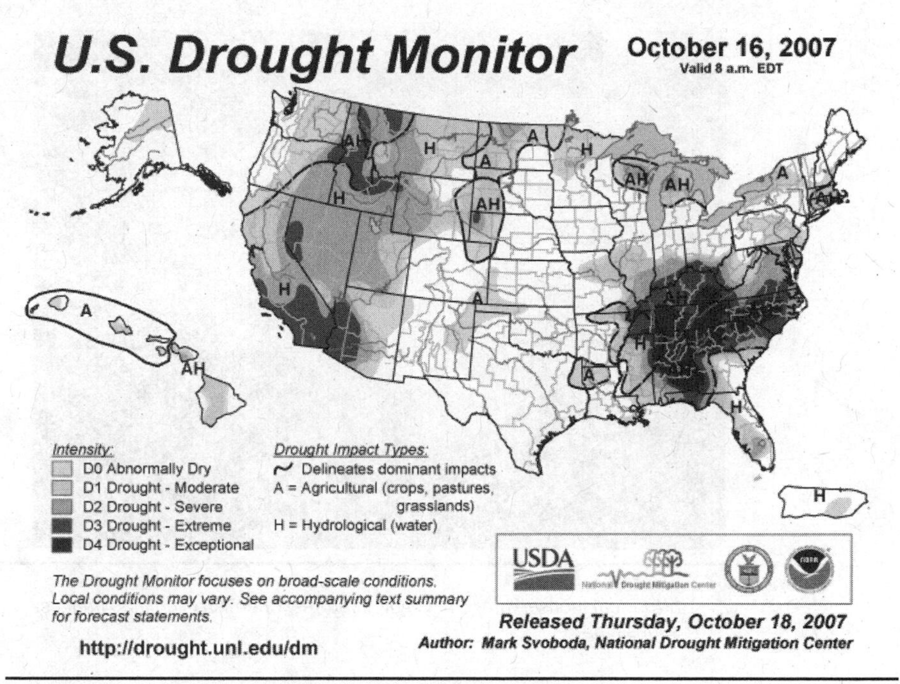

FIGURE 1-16 The desertification of the United States
http://www.crh.noaa.gov/images/pah/drmon_1016.gif

- Must be refined
- Produces waste products that must be disposed of
- Must be transported to the consumer

Fresh solar, on the other hand, offers us these advantages:

- No pollution or destruction of the land
- No heavy equipment required
- No fossil fuels required
- No refining or waste products
- Delivered direct to you for free
- Widely available
- Clean and abundant

So what are we waiting for? All types of solar energy are currently ready, available, and cost-effective. By reading this book, I hope you can learn how to make the most of solar energy now and for the future (Figure 1-17).

FIGURE 1-17 Sunrise
http://earthobservatory.nasa.gov/Features/ChemistrySunlight/Images/sunrise.jpg

CHAPTER 2

The Benefits and Detriments of Solar Energy

If you are reading this book, you either want to be informed about solar power or you want to use solar power in some capacity. Solar power is the paramount source of "alternative energy." (I place alternative energy in quotation marks because I believe that, within two generations, oil will become the alternative power source for mankind.)

Solar energy is unsurpassed by any other form of energy. The reasons for using solar power over any other form of energy are numerous:

- It is delivered everywhere, every day.
- It is completely free to use, collect, and store.
- It is free from transportation costs.
- It is free from pollution.
- It is an unlimited resource.
- It is the most reliable source of energy.

By comparing the advantages and disadvantages of solar power, you can see that solar has many advantages and few disadvantages. The advantages of solar power are obvious:

- Stable
- Reliable
- Abundant
- Free for everyone (after the costs of the equipment and setup)
- Available everywhere
- The most cost-efficient form of energy, if used correctly
- Clean power (zero emissions, zero health risks)
- Zero CO_2 emissions during use

- Preserves our natural environment, air quality, and limited water supplies
- No delivery costs
- No pollution
- Almost maintenance free
- Eligible for tax rebates when you set up solar panels
- You can profit from your extra solar energy
- Generates local jobs
- Provides greater energy security over the long term

The disadvantages of solar power pertain only to one form of solar energy—photovoltaic (PV) solar power (Figure 2-1):

- High initial cost
- Production of solar panels creates pollution
- The sun is not always available, so solar is an intermittent power source
- Disposal issues of old solar panels

The first two listed disadvantages of PV solar will soon be less important, as solar costs are decreasing (Figure 2-2). I am often verbally assaulted when I claim that solar energy is the most cost-effective form of

FIGURE 2-1 Photovoltaic solar panels
http://www.in.gov/oed/images/Bloomington-High-School-SouthWeb.jpg

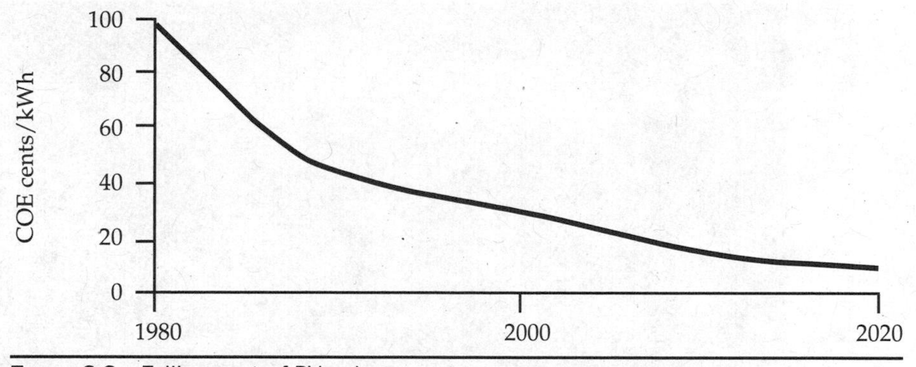

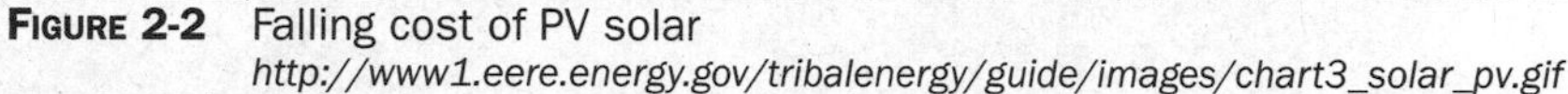

FIGURE 2-2 Falling cost of PV solar
http://www1.eere.energy.gov/tribalenergy/guide/images/chart3_solar_pv.gif

power, past, present and future. But this is not only my opinion—it's a fact. Individual solar projects can result in both personal and financial rewards in addition to cost savings.

In addition, the introduction of commercially available thin-film solar technology will change how solar energy is used (Figures 2-3 and 2-4). The promise of thin-film solar is a dramatic reduction in cost and speed of production, as thin-film solar's retail purchase price is 80 percent less than that of current PV solar panels.

The Solar Power Convergence

If you have chosen to use solar power now, you should be congratulated, because now is the perfect time to make use of this energy source. Welcome to the convergence.

FIGURE 2-3 Thin-film solar
http://www.nrel.gov/data/pix/Jpegs/15779.jpg

Figure 2-4 Thin-film solar variation
http://www.pnl.gov/news/images/photos/20090723100015409.jpg

In business, when a convergence occurs, fortunes are made, as many independent factors unite to create an opportunity of unprecedented proportion. The dramatic rise in oil prices, global warming, increased energy demand, traditional inefficient home building, and many other factors are converging to create unparalleled opportunities for alternative energy sources. You can participate in this convergence and therefore benefit personally.

In this convergence, renewable energy (Figure 2-5) will eventually replace traditional, nonrenewable forms of energy.

You'll remember from Chapter 1 that the sun produces and delivers more energy in one hour than the world uses in one year. Figure 2-6 shows that the most populous areas in the world receive solar energy both efficiently and effectively. Solar energy is a technically viable option worldwide, but today's initial cost is prohibitive for most of the world's people.

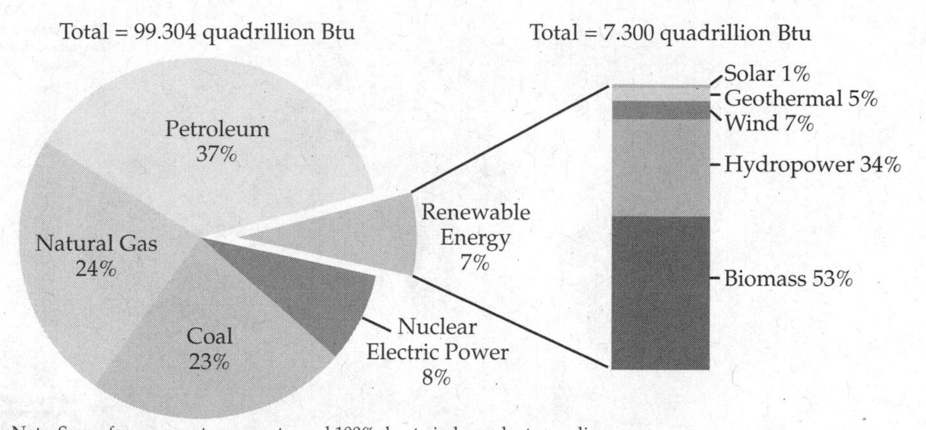

Note: Sum of components may not equal 100% due to independent rounding
Source: U.S. Energy Information Administration, *Annual Energy Review 2009*, Table 1.3, Primary Energy Consumption by Energy Source, 1949–2008 (June 2009).

FIGURE 2-5 Roles of renewable energy in the U.S. energy supply (2008)
http://tonto.eia.doe.gov/energyexplained/images/charts/role_of_renewables_in_us_energy-large.png

The current total world electricity consumption is approximately 17.5 trillion kilowatt-hours (kWh) and is increasing, as shown here.

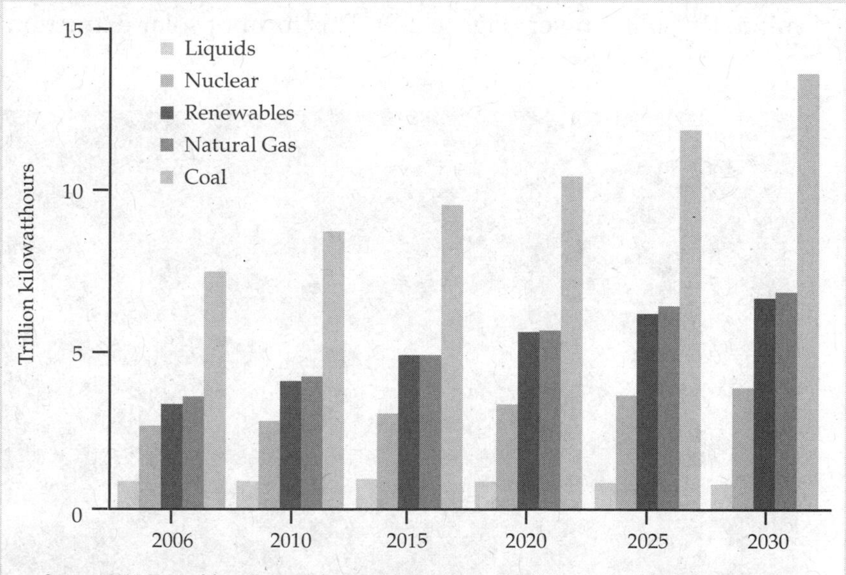

Sources: 2006: Derived from Energy Information Administration (EIA), *International Energy Annual 2006* (June–December 2008), web site www.eia.doe.gov/iea. Projections: EIA, World Energy Projections Plus (2009).

World electricity usage
http://www.eia.doe.gov/oiaf/ieo/images/figure_51small.jpg

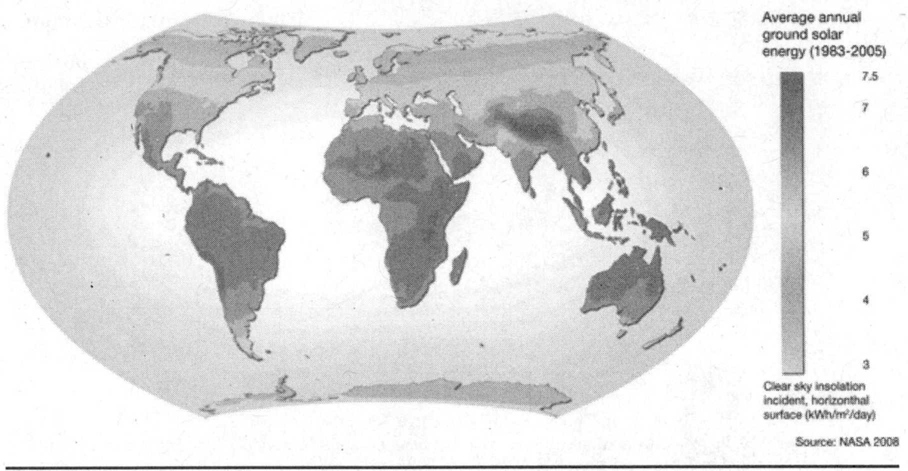

FIGURE 2-6 Average annual ground solar energy, 1983–2005
http://tonto.eia.doe.gov/energyexplained/images/charts/world-solar-large.gif

Types of Solar Power

Today, when we think of solar power, one form of solar usually comes to mind: PV solar power (Figure 2-7). This form of solar extraction is what

FIGURE 2-7 Solar panels
http://www.ia.nrcs.usda.gov/news/images/Pics/solarpanels4621.gif

most people call solar power, but PV solar power (changing solar radiation to electricity) is only one form of solar energy.

Other forms of solar energy include the following:

- Passive solar (Figure 2-8)
- Solar heating (Figures 2-9 and 2-10)
- Solar lighting (Figure 2-11)
- Indirect solar energy (Figure 2-12)

FIGURE 2-8 Passive and PV integrated home
http://www.nrel.gov/data/pix/Jpegs/11672.jpg

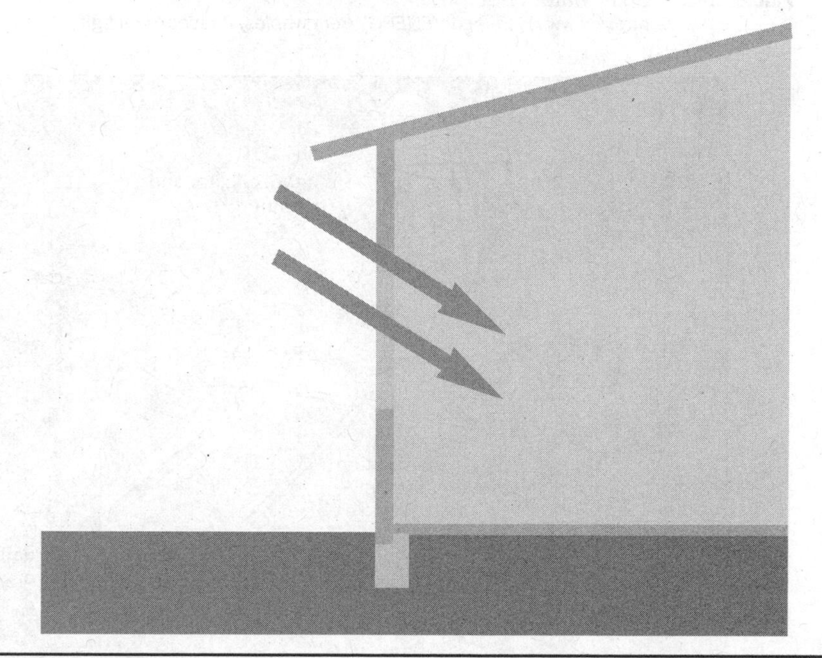

FIGURE 2-9 Solar light heat
http://resourcecenter.pnl.gov/cocoon/morf/ResourceCenter/dbimages/full/11.jpg

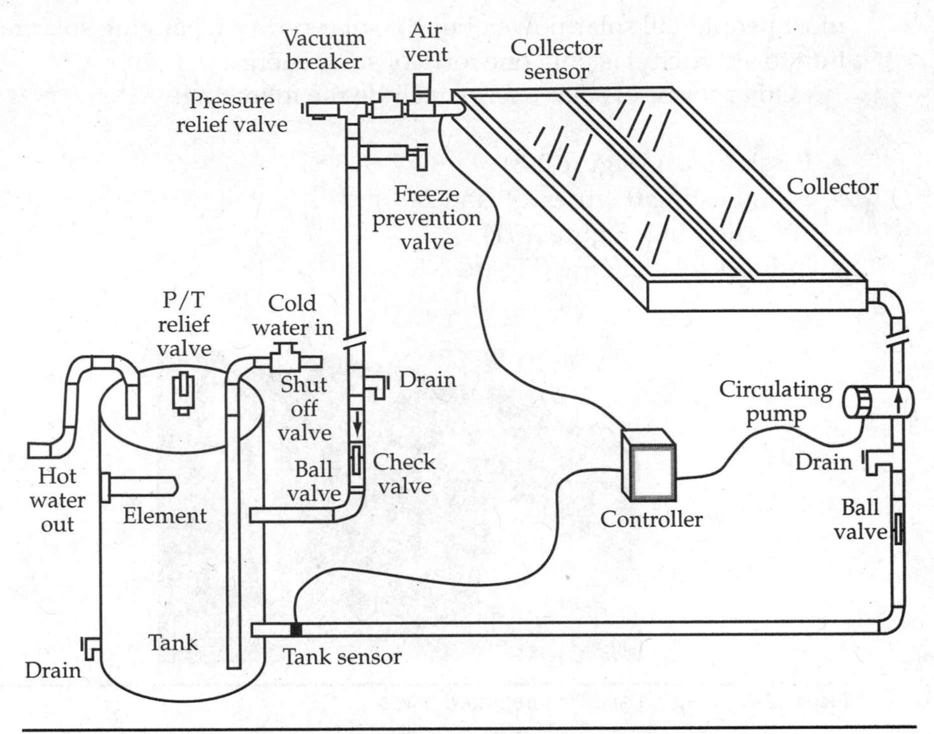

FIGURE 2-10 Solar water heat
http://www.dnr.mo.gov/ENERGY/renewables/images/direct.gif

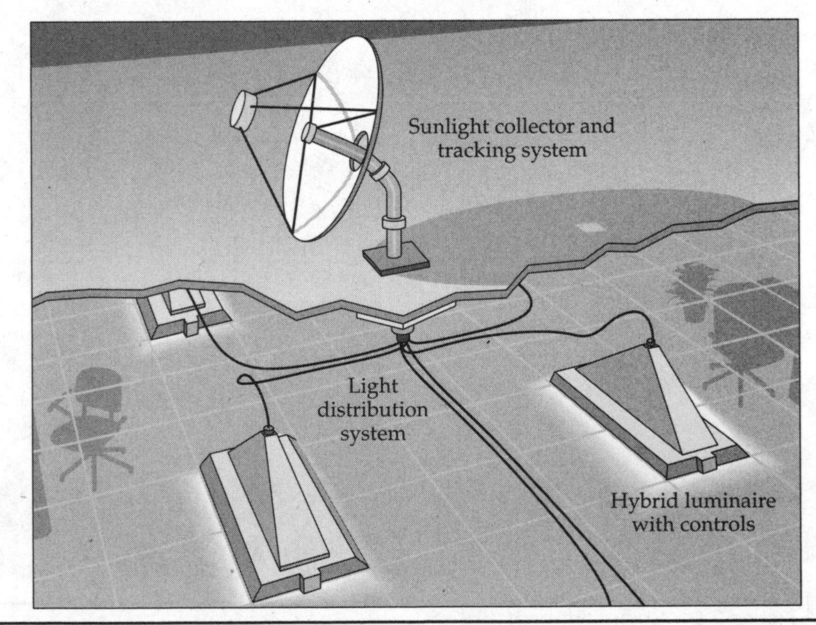

FIGURE 2-11 Solar lighting
http://www.ornl.gov/info/ornlreview/v33_1_00/p22.jpg

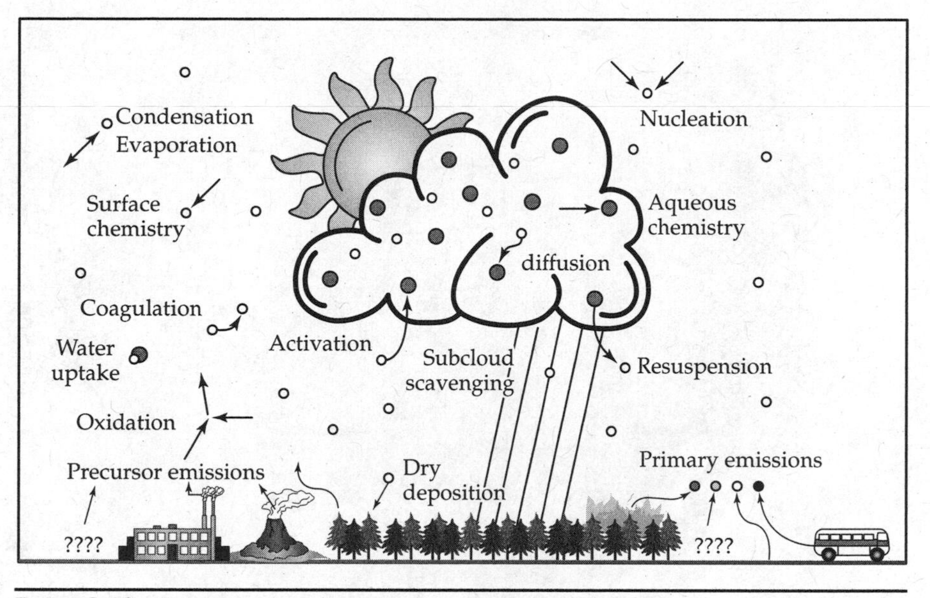

FIGURE 2-12 Indirect solar interactions
http://www.pnl.gov/atmospheric/research/aci/images/aerosol_clouds.jpg

Passive solar is a form of energy that you already use; it's covered extensively in Chapter 5. In the dead of winter, for example, even on an extremely cold day, I can open the blinds on the south-facing windows of my home, and this simple adjustment will increase the temperature in my rooms by 2.4°C or 5°F—just by opening the blinds. It doesn't get any simpler or cost effective than this.

Throughout the United States, solar power can be used as a primary source of energy during all four seasons (Figure 2-13).

Other Forms of Energy

Three other viable forms of energy are available on our planet: wind, geothermal, and tidal power.

Wind Power

Wind, an all-but-forgotten form of energy, is neither new nor revolutionary. Wind has powered sailing vessels for at least 5,500 years. Windmills have a history of at least 2,500 years, and have been used to grind grain, pump water, and create mechanical assistance in manufacturing. In 1927, Joe and Marcellus Jacobs produced and sold wind turbines with genera-

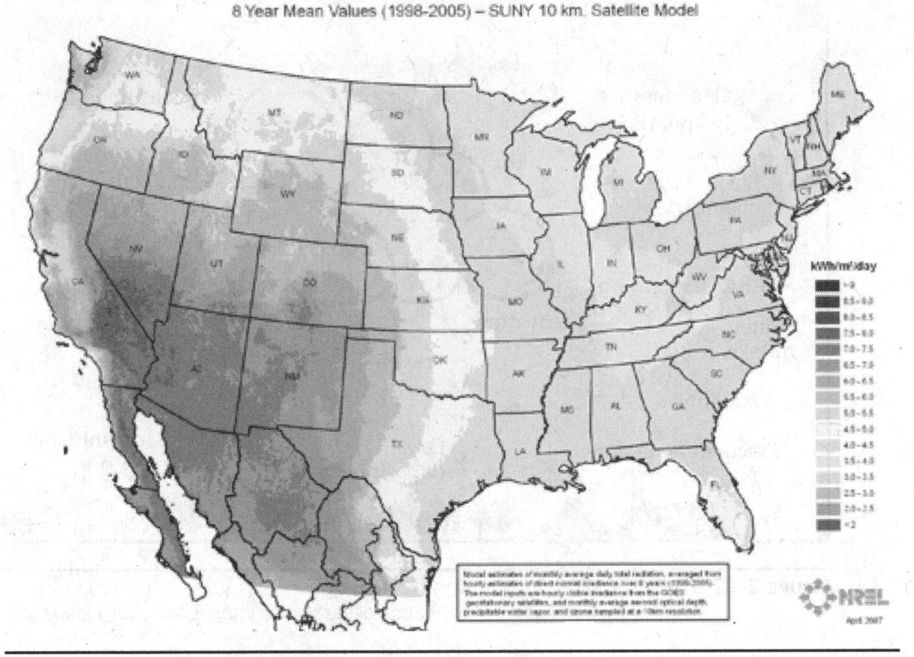

FIGURE 2-13 Annual direct normal solar radiation
http://www.nrel.gov/csp/troughnet/images/map_normal_radiation.gif

tors for use on farms to pump water from wells to be used in the farmhouse as well as in the fields and barns.

Truth is, wind power is solar energy. Air is set in motion by the pressure gradient force, a flow from high pressure to low pressure, a direct result of the warming and cooling of air by the sun. Simply put, when cold, dense air is placed next to warm, less dense air, wind results as nature tries to balance the pressure differences at each level in the atmosphere between the two air masses.

Wind is affordable but not accessible everywhere. A modern wind turbine requires a constant speed of about 33 mph, and although sustained wind speed is available in many areas, wind energy is not always as easily available as solar energy.

Geothermal Energy

Geothermal power has been used for at least 2.5 million years. Geothermal energy originates from the Earth's core. Because it is available only where tectonic plates converge (Figure 2-14), geothermal can supplement green energy but, like wind energy, it is not available everywhere.

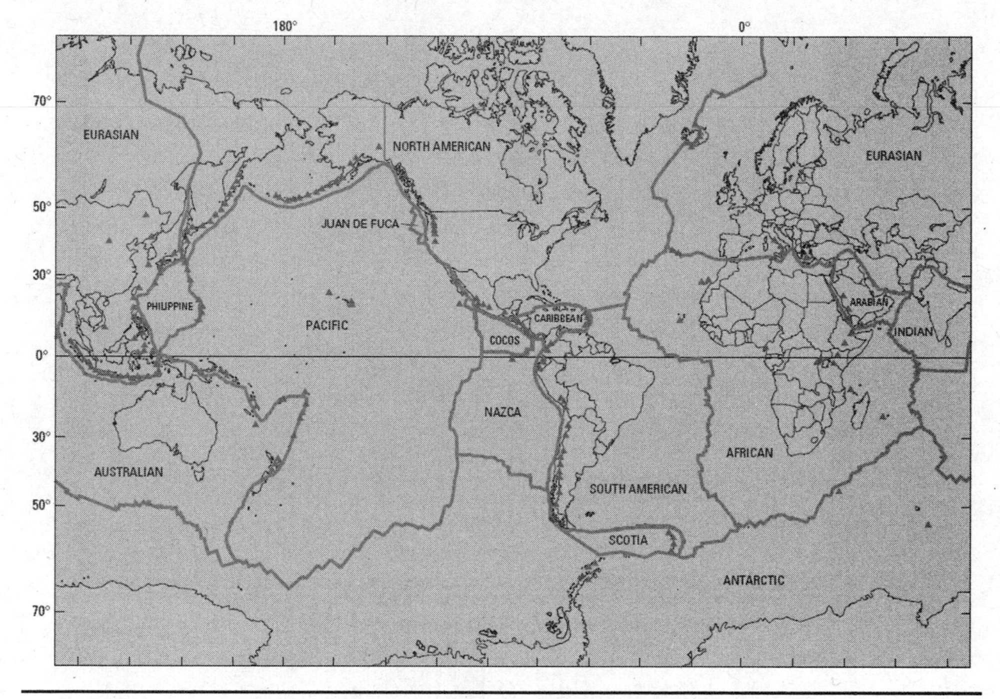

FIGURE 2-14 Geothermal world map (geothermal areas shown as triangles)
http://energy.usgs.gov/images/geothermal/geothermal_lithoplatesLG.gif

Tidal Energy

Another clean energy option available today, tidal power is not well known because it is invisible to most of us and quite expensive. As with most renewable technologies, tidal power is not new; in fact, it is one of the largest and oldest forms of energy used by man. Tidal mills were used in all western nations since 700 A.D. The British, French, Spanish, and Americans used tide mills as food storage ponds that were filled by the incoming tide: water was captured and then released through a water wheel, and the water wheel produced mechanical energy that provided power to mill the grain. One such mill in New York worked well into the 20th century.

The benefits of tidal power are numerous. Tidal turbines (Figure 2-15) are similar to wind turbines and are a completely clean source of energy. Tidal power is reliable, predictable, and constant, unlike wind. There is more energy potential in tidal power than in all of the oil in the world. If the offshore tidal power in the eastern United States were harnessed, it could power the entire world. The main detriment to tidal power is that it is currently one of the most expensive forms of energy to produce.

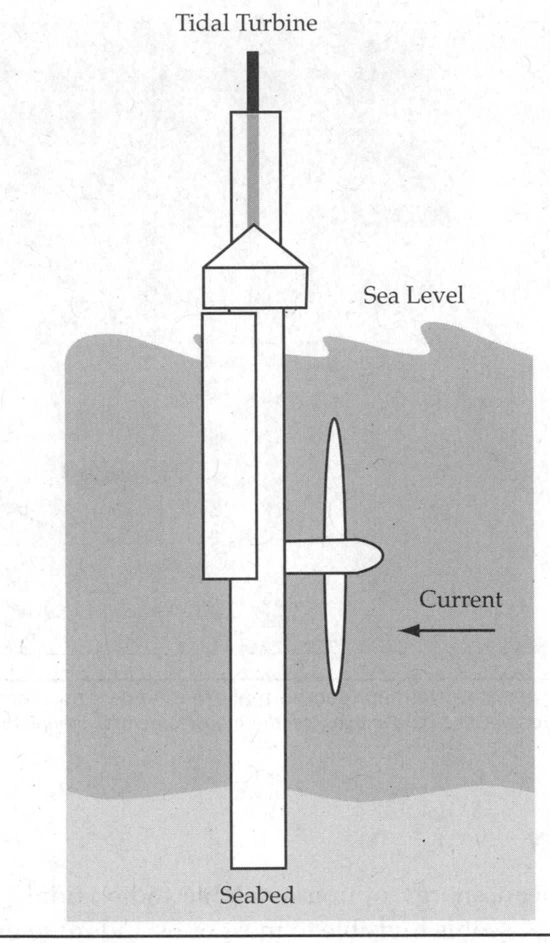

FIGURE 2-15 Tidal turbine
http://tonto.eia.doe.gov/energyexplained/images/tidalturbine.gif

Ultimate Benefits of Solar Energy

Everyone wants to be able to create a pleasant environment in their own personal space. Today, however, more than one half of the world's population cannot control their personal environment.

Comforts of a modern home include heating and cooling systems. The following devices account for the largest expenses you incur in your home but represent only some of the electrical devices that most people in developed nations presently use:

- Large electrical appliances, such as washer, dryer, refrigerator, stove, oven, lights, telephone, vacuum, television, stereo, dehumidifier, microwave, toaster oven, computer, media system, pool pump, sump pump, water filter, and much more.

- Small electrical appliances, such as shaver, hair dryer, coffee maker, blender, printer, scanner, copier, fax, shredder, chargers, tools, fans, and much more.

So what are all of these amenities and appliances costing us each day? Let's look at a few of the largest energy consumers in your home in the simplest way. The heating, cooling, and major appliances in your home account for the majority of your energy usage. What do you pay for these comforts? And what does each of these luxuries have in common? They all use electricity. In assessing your energy use (Figure 2-16), you should consider not only comfort controls, but electrical appliances. An evaluation of your home's energy use might motivate you to purchase solar power instead of conventional sources of power.

You now know solar energy comes in many forms, it is accessible, and can be cost effective. What other benefits does solar energy have? From a global perspective, solar will assist with the reduction of greenhouse gases (Figure 2-17).

Why we should all be using solar and other alternative forms of energy now is shown in Figure 2-18.

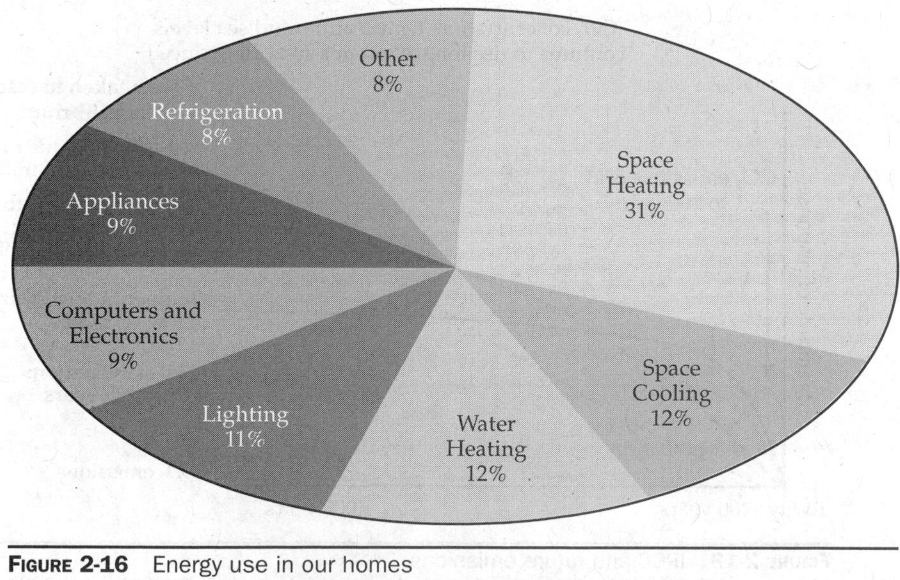

FIGURE 2-16 Energy use in our homes
http://www1.eere.energy.gov/consumer/tips/home_energy.html

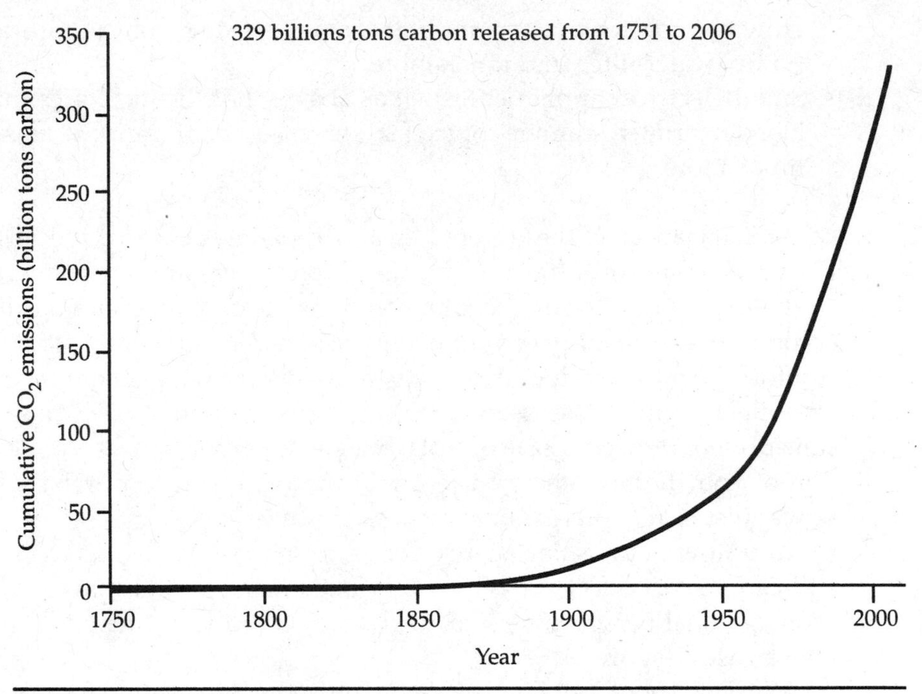

FIGURE 2-17 CO_2 fossil-fuel emissions
http://cdiac.ornl.gov/trends/emis/graphics/global_cumulative_1751_2006.jpg

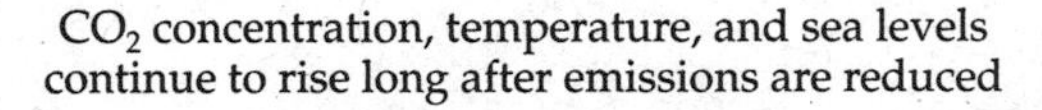

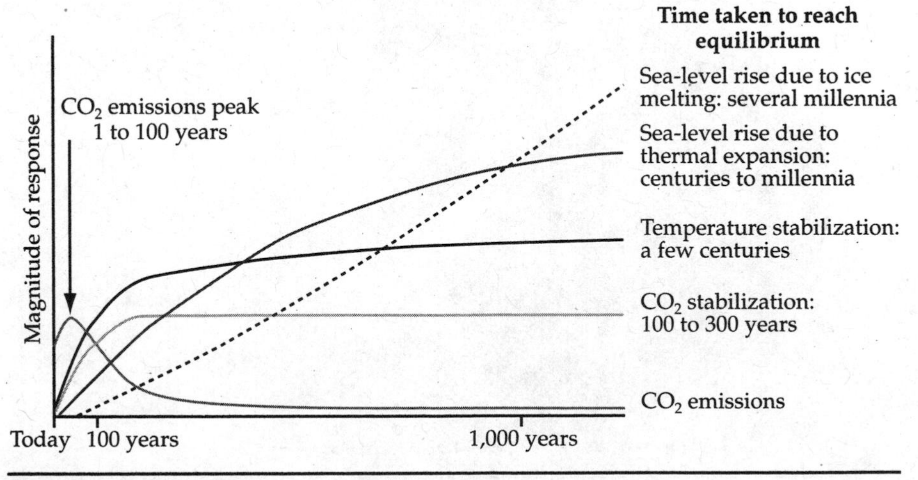

FIGURE 2-18 IPCC and future emissions
http://www.ipcc.ch/pdf/supporting-material/sectoral-economic-costs-2000.pdf

Challenges of Solar Energy

I saved my discussion of the challenges of solar to close this chapter. Earlier, I mentioned four challenges for solar energy:

- High initial cost
- Pollutants during the production of solar panels
- The sun is not always available
- Disposal of old solar panels

But solutions to these issues are available:

- Integrating all types of solar and taking advantage of loans, grants, and rebates can make solar a more cost-effective alternative.
- The pollutants from solar panel production are being reduced as technology changes.
- Solar has been shown to be effective in most populated areas of the world.
- The final issue, disposal, has not yet been addressed. However, science and big business are now researching profitable ways you can recycle old solar components.

CHAPTER 3
Types of Solar Energy

Today, in our homes and at work, most of us use many energy-intensive processes that make our lives easier. Drinking water is available simply by turning a spigot or tap, sewage and waste are safely removed and treated, and we use lights, heat, and automatic washers and dryers. These are everyday luxuries brought about by energy-intensive processes. Fact is, we could live without them all, but we choose not to.

The energy required to power these technologies is traditionally based on fossil fuel consumption. However, today's technology has advanced to the point at which we now have alternative options for how we supply energy for our luxuries. We have a viable and important energy source in solar energy.

Solar energy is the only choice for most of the world. The most efficient form of energy, it can and should be used to save or conserve energy or as a complementary technology to more traditional energy sources.

Energy and Oil

The world's dramatic increase in energy consumption began in the late 1800s with the mass consumption of oil. Petroleum was refined to produce kerosene, which replaced whale oil, primarily for lighting. Petroleum byproducts include kerosene, gasoline, and other oil products (as shown in Figure 3-1).

Coal was used for home heating until the 1950s, and crude oil is still in use in some homes. The massive use of oil began during World War II with the requirement of transportable fuel for mechanized warfare. The WW II generation came home from the war to a new way of life. Figure

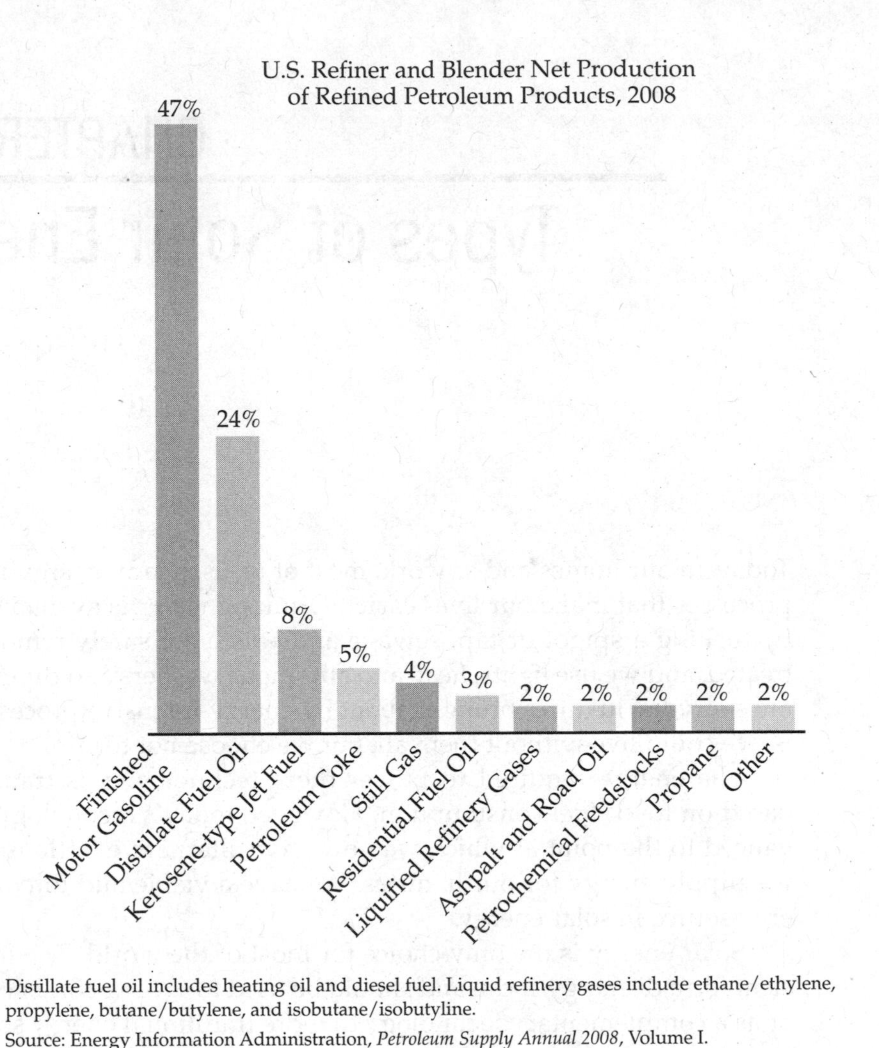

FIGURE 3-1 Petroleum byproducts
http://tonto.eia.doe.gov/energyexplained/images/charts/us_refiner_blender_net_production_refined_petroleum_products-large.gif

3-2 shows homes in Levittown, New York, built from 1947 to 1951 and considered the first truly mass-produced suburb. Levittown is regarded as the archetype for postwar suburbs throughout the United States. Subdivisions of the era offered a home for every family, a car in every driveway, and all the modern amenities. Luxuries formerly unimaginable were available to the masses.

All of suburbia required electrical energy, then and today produced by coal and oil. As suburbs increased nationwide, so did the consumption of energy (Figure 3-3).

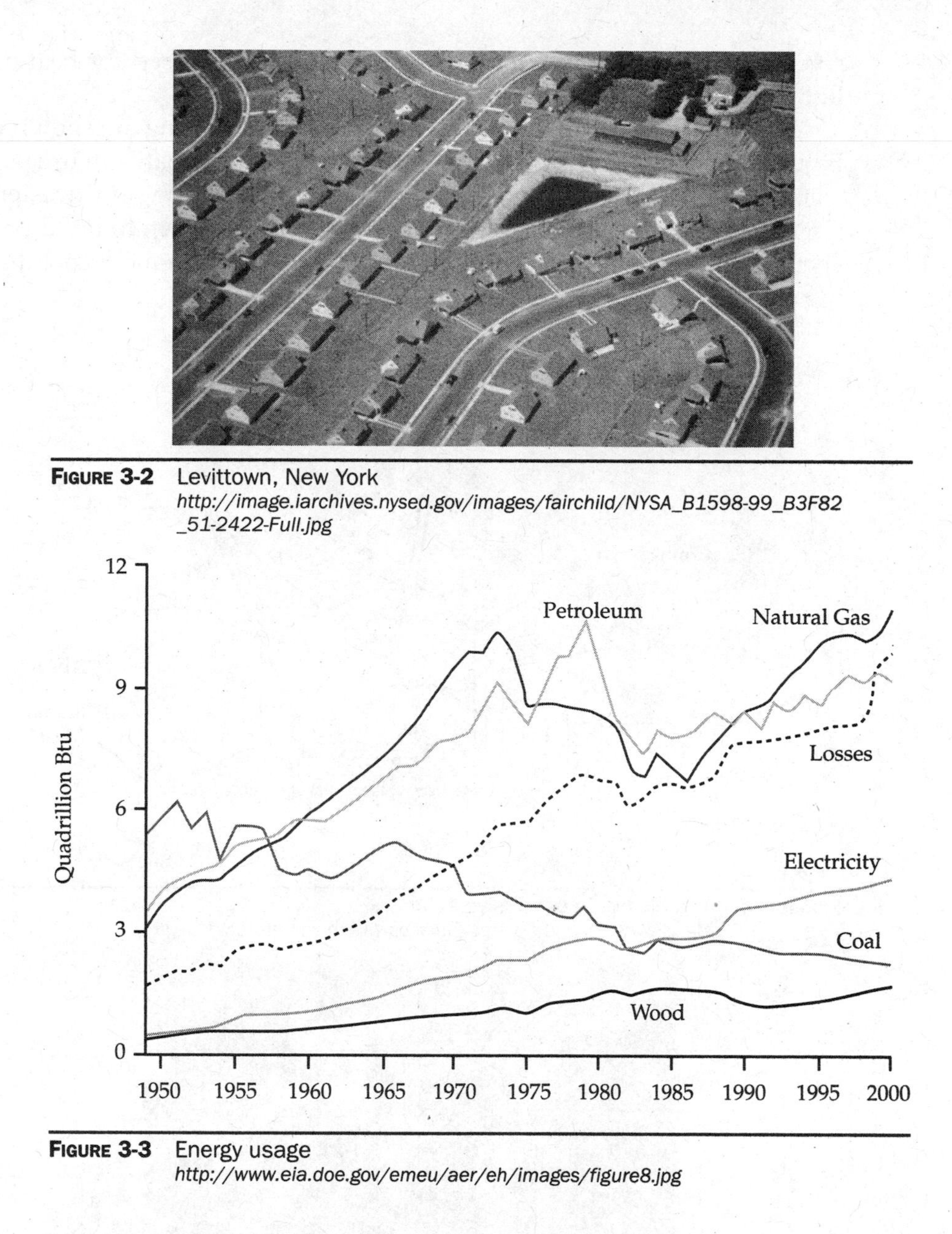

FIGURE 3-2 Levittown, New York
http://image.iarchives.nysed.gov/images/fairchild/NYSA_B1598-99_B3F82_51-2422-Full.jpg

FIGURE 3-3 Energy usage
http://www.eia.doe.gov/emeu/aer/eh/images/figure8.jpg

Passive Solar

Passive solar is far more available, affordable, and earth friendly than traditional energy sources. Passive solar involves the direct use of sunlight without the need for any type of mechanical assistance. Passive solar energy can be used directly or in combination with other energy forms. (Passive solar is covered in depth in Chapter 5.) It can be used for heating, lighting, cooling, water purification, and much more.

The five elements of passive solar heating and lighting are shown in Figure 3-4. A properly designed home in a temperate climate can use passive solar heating without additional heat sources. In the southwestern United States, for example, many individuals have chosen to build passive solar homes, as shown in Figure 3-5. These homes remain cool during a 110°F day and warm even during a 30°F evening.

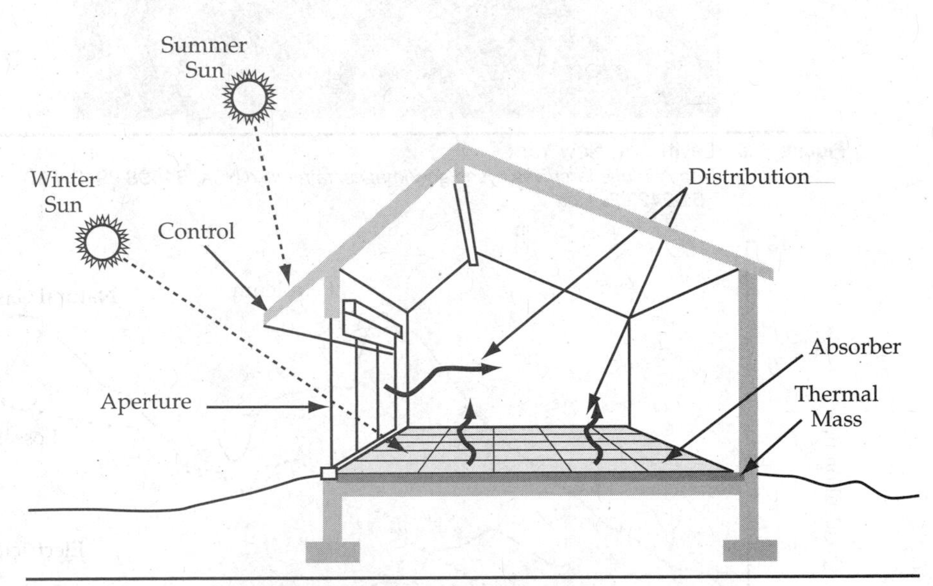

FIGURE 3-4 Five elements of passive solar
http://www.energysavers.gov/images/five_elements_passive.gif

FIGURE 3-5 Passive solar home
http://www.gilacountyaz.gov/DepartmentFiles/CommunityDevelopment/earthship.jpg

Passive solar is the easiest form of energy to use in the home for the following reasons:

- Passive solar is free to use.
- Solar energy is available everywhere the sun shines.
- The energy is delivered free to your home.
- Passive solar creates zero pollutants.
- Passive solar can be used on a consumer, commercial, or industrial scale.
- Passive solar has a 4.5-billion-year history of reliability.

Passive solar can be used in combination with mechanical devices, which, once installed, require little or no additional energy. These devices can replace others that require large amounts of energy. A perfect example is a solar hot water heater (Figure 3-6).

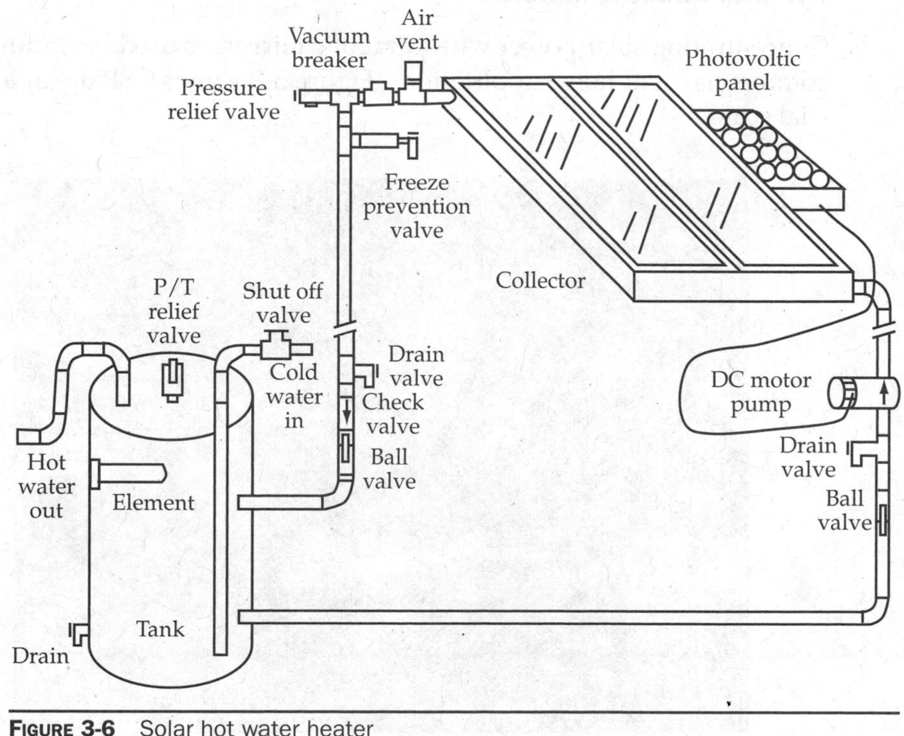

FIGURE 3-6 Solar hot water heater
http://www.westnorfolk.gov.uk/images/Solar%20Hot%20Water%20Heating%20Diagram.gif

Concentrated Solar Power

Concentrated solar power (CSP) systems use focal material, such as lenses or mirrors, and a tracking system to focus sunlight onto a small area to produce intense heat. The concentrated light is then used to create heat as a source for conventional power.

CSP is not a new technology. In the 1860s, French inventor Augustin Mouchot created a solar-powered engine that converting solar energy into mechanical work via steam power (Figure 3-7). Mouchot believed that coal, the fuel of the industrial revolution, was in limited supply and would eventually be unavailable. Mouchot and his engine were victims of the same tribulations of today, however. More efficient delivery and use of coal in the 1870s allowed prices to drop, and Mouchot engines became economically unviable.

CSP is again being explored as a reasonable option for clean energy, with CSP photovoltaic (PV) panels installed in both home and industrial applications.

CSP with Parabolic Mirrors

Concentrating solar power with parabolic mirrors is used for industrial, commercial, and home applications. Figure 3-8 shows CSP on an industrial scale.

FIGURE 3-7 Mouchot solar-powered engine
http://en.wikipedia.org/wiki/File:Mouchot1878x.jpg

FIGURE 3-8 CSP parabolic
http://teeic.anl.gov/images/photos/PIX_01223_Trough_Face_Kramer.jpg

The interest in CSP is a result of the high cost of PV panels. Concentrating solar is simplicity in design. It employs reflective material, such as mirrors, polished metals, or even fabrics such as Mylar, deployed in a radius around a central tower. The reflective materials are remotely controlled to track the sun during the day.

The central tower on which the reflective materials are mounted (Figures 3-9 and 3-10) is filled with the material to be heated, which could be water, oil, salt, or other substances that can store light energy as heat. As the sun rises, it heats the water to more than 1000°F. The heated material creates steam and turns a turbine to power a generator, which produces electricity. The heated material will be at relative low temperature during the beginning of the day and near its hottest temperature by the end of the day. The system should theoretically store enough heat energy to continue to produce steam, and therefore, electricity day and night.

Concentrating solar power of this type is not a 24-hour-a-day solution, however. Although CSP plants are usually constructed in sunny areas of the world, even the great deserts sometimes have more than a day of cloud cover. Sandstorms, dust, birds, inclement weather, and intermittent sunlight all contribute to inefficiencies in this type of system. Some of these power stations are used to produce intermittent energy when the sun is available. When a constant power solution is required, natural gas or oil is usually substituted to maintain a constant temperature.

With many variations of concentrated solar power available, you will see more CSP solutions in the future.

Stirling Engine

A variation on concentrated solar power is the Stirling engine, shown in Figure 3-11. Like CSP with parabolic, this is concentrated solar power;

FIGURE 3-9 Central focal point
http://teeic.anl.gov/er/solar/restech/desc/index.cfm

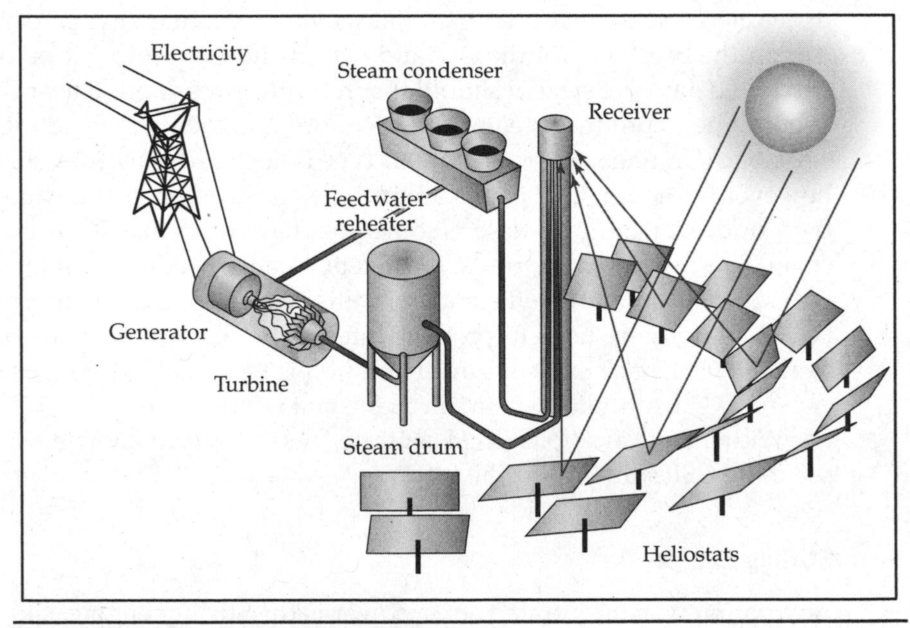

FIGURE 3-10 How CSP mirrors work
http://www1.eere.energy.gov/solar/images/power_tower.jpg

however, instead of focusing the sunlight on a central tower, the sunlight is focused on an "engine" to create mechanical work and then electricity. This type of energy production has many benefits but one challenge that separates it from other forms of concentrated solar energy: The Stirling engine produces a loud sound similar to that of an internal combustion engine, which prohibits its use inside the home.

The Stirling engine functions on the principle of a differentiation in temperature. You may be familiar with the Stirling engine from a small engine used in high school science class (Figure 3-12). A teacher may have held this engine on his or her hand, or over a cup of hot water, and the temperature variation would turn the wheel on top.

This is similar to what occurs with the larger solar version of the engine (Figure 3-13). The concentrator focuses the sunlight on the receiver, or the Stirling engine, and produces usable energy. This system is then tied into the grid for some type of immediate power usage (Figure 3-14).

The Solar Energy Development Programmatic EIS is an excellent reference for solar energy of all types. Visit http://solareis.anl.gov/guide/photos/index.cfm to see the many different types of concentrated solar energy with images. This website demonstrates current and recent developments in solar technology.

FIGURE 3-11 Concentrated solar power Stirling engine
http://www.sandia.gov/news/resources/news_releases/images/2009/stirling.jpg

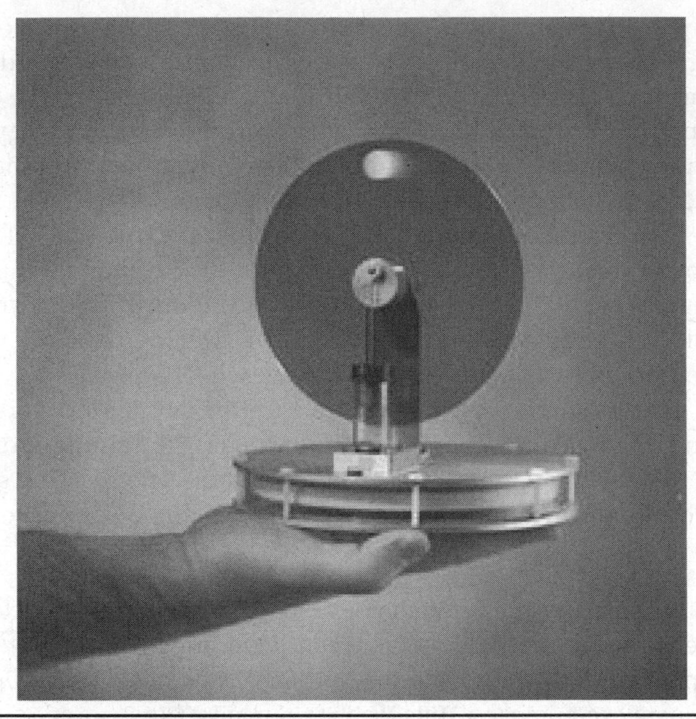

Figure 3-12 Handheld Stirling engine
http://grcimagenet.grc.nasa.gov/grcdigitalimages/1996/1996_00555L.jpg

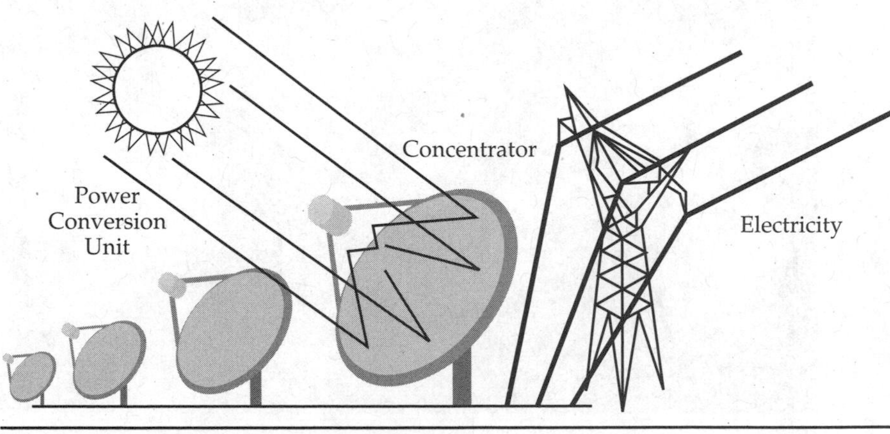

Figure 3-13 Parts of the solar Stirling engine
http://www1.eere.energy.gov/solar/dish_engines.html

Figure 3-14 Stirling engine
http://www.srpnet.com/newsroom/graphics/tesserasolar/January2010/PC230480_Large.jpg

Figure 3-15 Concentrating and using energy from the Stirling system
http://www.sandia.gov/news/resources/news_releases/images/2009/stirling.jpg

Solar Water Heaters

A traditional type of solar water heater can heat water to a temperature of 130° to 180°F. The evacuated tube collectors (Figure 3-16) are the most efficient collectors available. These collectors function similar to a thermos: A glass or metal tube containing the water or heat transfer fluid, antifreeze, is surrounded by a larger glass tube. The space between the tubes is a vacuum, so very little heat is lost.

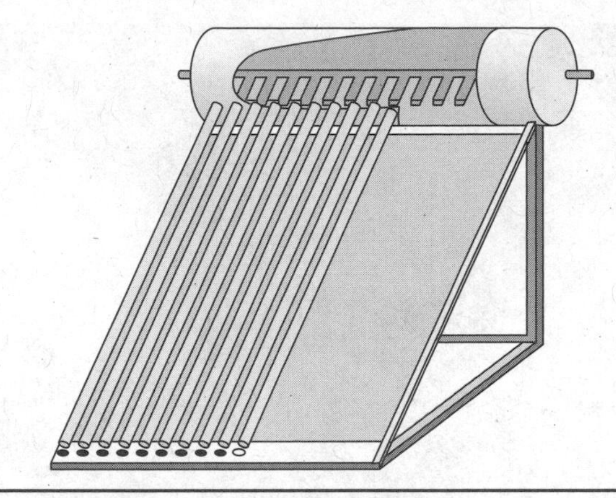

FIGURE 3-16 Evacuated tube solar hot water heater
http://www.energystar.gov/ia/products/fap/images/products/wse_med.png

A similar type of CSP can also be employed for home use. A parabolic solar water heater, shown in Figure 3-17, uses a parabolic mirror(s) to heat water to 400°F or higher. At these temperatures, the hot water can be used to supply heat for homes set up with liquid heating systems, such as baseboard systems heating. In these solar systems, a parabolic reflector or mirror focuses sunlight on a center focal point filled with liquid. The liquid is heated and pumped into the home. The heat is released into the home and is returned to the solar concentration point to be reheated.

FIGURE 3-17 Parabolic solar water heater
http://www.nrel.gov/data/pix/Jpegs/16142.jpg

Updraft Solar

A related solar technology is updraft, or chimney, solar shown in Figure 3-18. With updraft solar, a large surface area of the ground is covered in

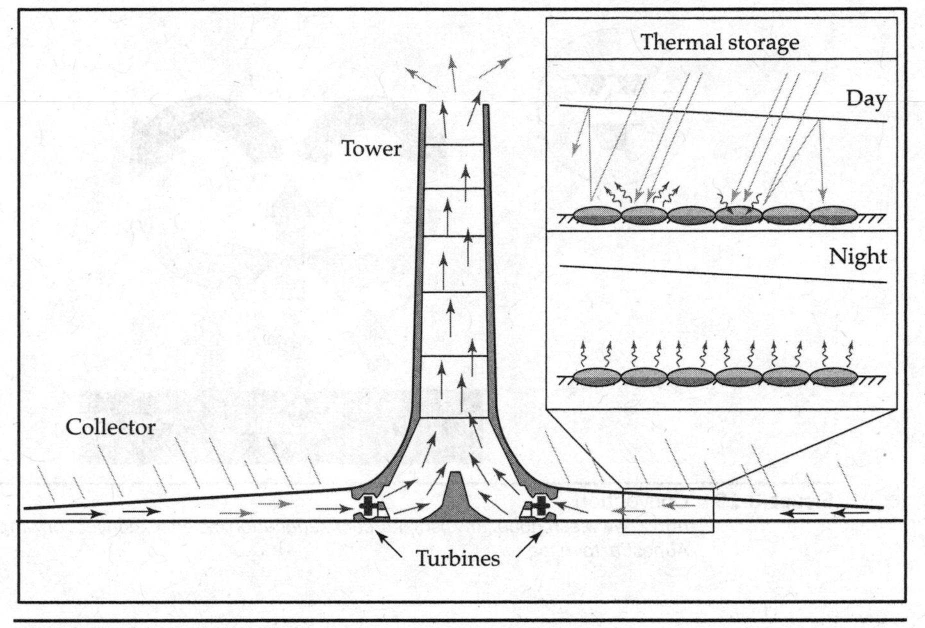

FIGURE 3-18 Updraft solar
http://en.wikipedia.org/wiki/File:Solar_updraft_tower.svg

a radial area around a tower. The covered area is open at the edges of the radius and slopes upward toward the tower. Natural convection (Figure 3-19) allows the air to flow into the tower and turn turbines at the base of the tower. The hot air rises and naturally flows out of the top of the central tower, and the process continues as long as a differential in temperature exists. This form of solar energy is of the concentrating type. Sunlight is used to heat air and concentrate that energy on the turbines.

Concentrating PV Solar

Concentrating solar is also available in photovoltaic form (Figure 3-20). CPV solar functions similar to traditional PV solar, except a lens is used to focus sunlight on a single photo cell instead of an array of cells. The most expensive part of PV solar is the photo cell. By reducing the amount of cells required, the cost of CPV solar is reduced dramatically. The exciting element of focused or concentrating PV solar is that it can be used anywhere and has all of the benefits of traditional PV at a fraction of the cost.

Concentrating and updraft solar are primarily commercial- and industrial-scale operations.

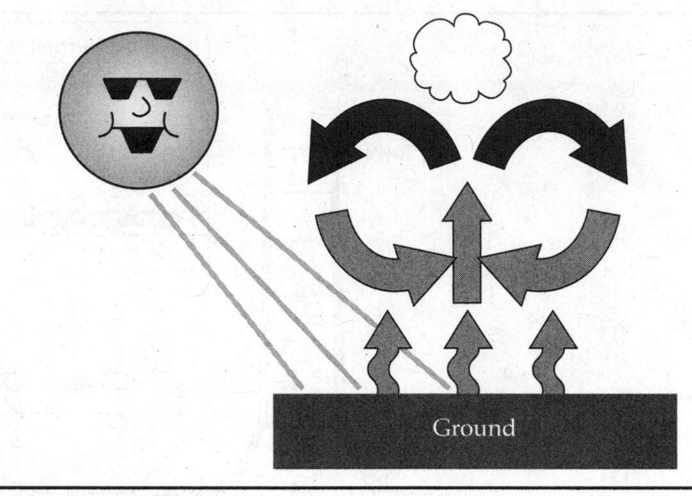

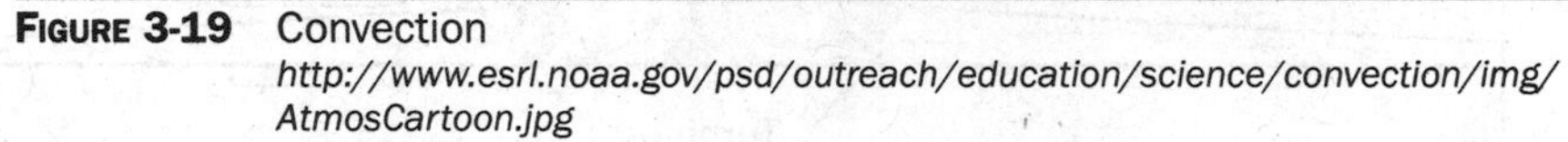

FIGURE 3-19 Convection
http://www.esrl.noaa.gov/psd/outreach/education/science/convection/img/AtmosCartoon.jpg

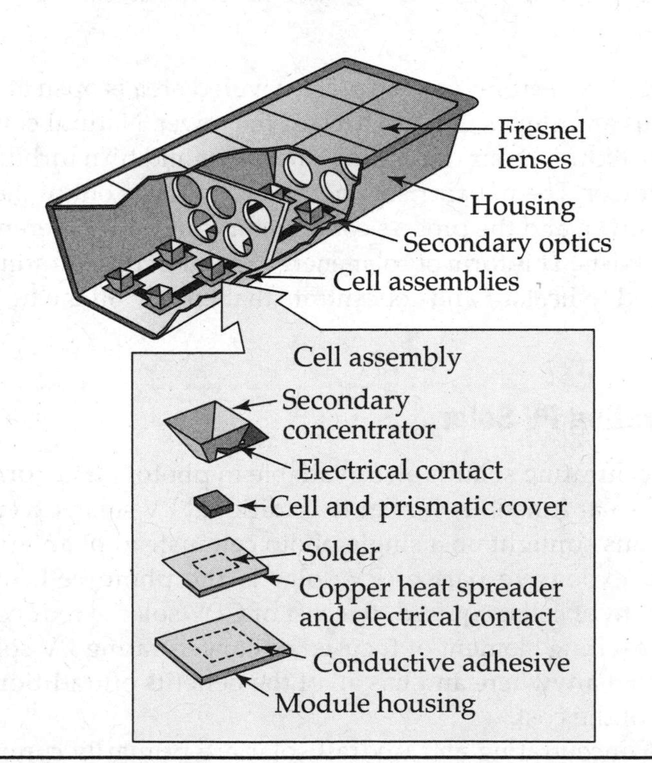

FIGURE 3-20 Concentrating PV
http://www1.eere.energy.gov/solar/images/illust_concentrator.gif

The National Renewable Energy Laboratories offers information on its website about current and emerging technologies at http://www.nrel.gov/csp/lab_capabilities.html#aom_lab.

These technologies can also be merged with fossil fuels to create a constant supply of power. The addition of a battery or fuel cell systems can make these solar units independent and a constant source of power. Connecting to the power grid is also an option (Figure 3-21).

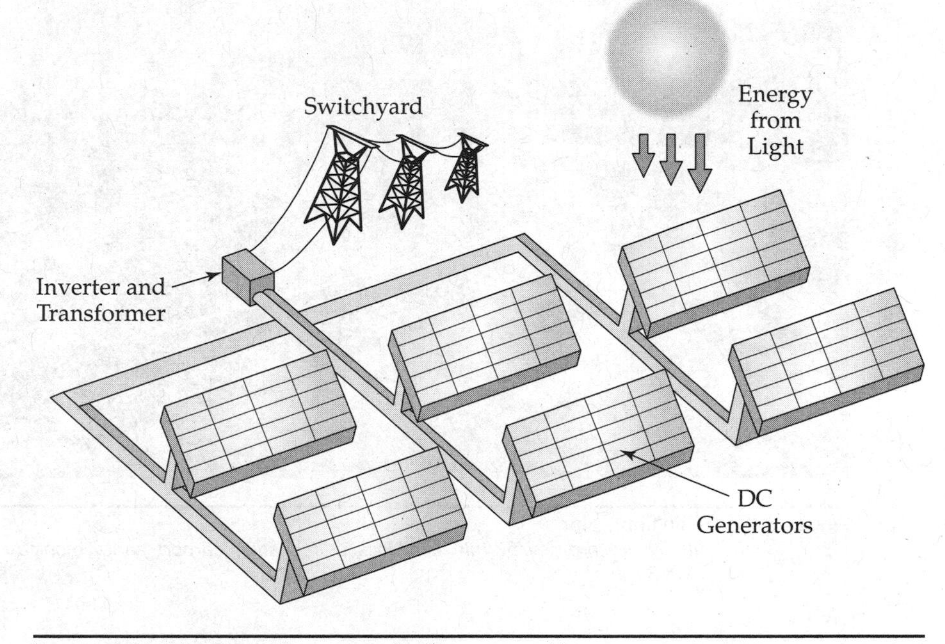

FIGURE 3-21 Adding solar energy to the power grid
http://www.tva.gov/greenpowerswitch/images/diagram_solar.gif

Types and Benefits of PV Solar

PV solar is being promoted for homeowners and business all over the world. PV solar offers an immediate and direct replacement for fossil fuel energy. The energy we need can be provided immediately with solar and used onsite. If you produce more electricity than you require, you can provide power to your local utility.

Thin-Film Solar

Traditional PV panels are the most expensive form of PV solar but also the most efficient. Thin-film solar, such as that shown on a California carport roof in Figure 3-22, functions like traditional PV panels but is less efficient. Thin-film panels are significantly less expensive and can be formed into other surfaces. Thin film is also more durable and can be used as a portable charger or power source.

FIGURE 3-22 Thin-film solar
http://www.nrel.gov/pv/thin_film/docs/united_solar_carport_santa_monica_2003.jpg

Focused-Lens and Specialty PV

Focused-lens PV can use as much as 100 times less silicon than traditional solar PV cells, which is the most costly ingredient in the manufacture of solar panels.

Lenses mounted on top of a photovoltaic cell replace multiple cells in an array (Figure 3-23)—one cell per module instead of multiple photo cells.

Small device PV cells can be traditional or thin-film solar panels. Small solar devices can be used for recreational vehicles, portable radios, or other specialty devices (Figure 3-24). Small solar devices are becoming popular, and specialty devices are being designed to function using solar energy exclusively. As the efficiency of thin-film solar increases, more applications will emerge and be integrated into everyday items.

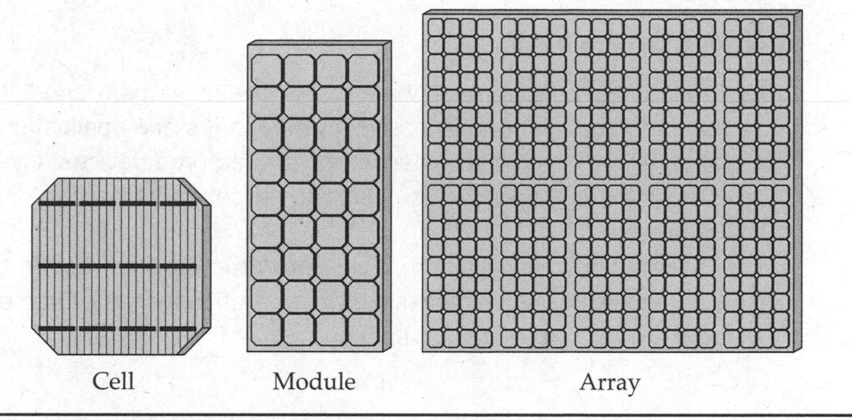

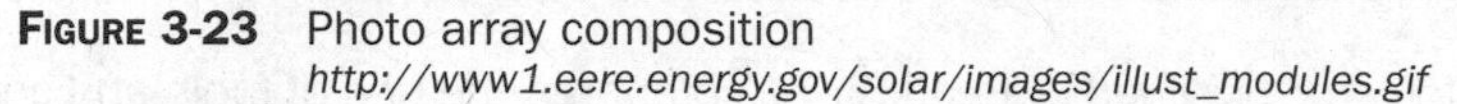

FIGURE 3-23 Photo array composition
http://www1.eere.energy.gov/solar/images/illust_modules.gif

FIGURE 3-24 Use of PV cells in research
http://www.nrel.gov/pv/thin_film/docs/gse_30_w_cis_antarctic2005.jpg

Solar Costs

The year 2010 is a pivotal year for solar energy. With companies such as First Solar (http://www.firstsolar.com/en/index.php) claiming to offer

Beyond Passive Solar

Two main types of solar energy are readily available: heat and light. Yet, as you can see, solar energy has many categories and applications, and not all of it requires the assistance of manufactured devices, lenses, or mirrors. The sun works for us every day—I call this "beyond" passive solar energy.

The amount of energy essential to grow the food we require to feed 7 billion people would be impossible by any artificial means. Even our fossil fuels are remnants of solar energy.

This calculator, provided by Penn State, compares the use of fuels and costs of various types of fuel. Try it out—you might be surprised by the results.

http://energy.cas.psu.edu/EnergySelector.html

remarkably inexpensive energy generated by solar panels, we may see a revolution in PV solar electricity.

The Department of Energy website (http://www.energy.gov/energysources/solar.htm) provides excellent discussions of all potential energy sources. Every available statistic and type of energy is listed and accurate. The measurements and comparisons of energy help you to answer the question, What will solar energy cost me?

Today, how you buy or use energy is limited only by your imagination. For the first time in industrialized history, you can become your own power plant (Figure 3-25). You can create and add energy to the power grid. You can design a home that requires little or no energy, because solar is delivered free to your home every day.

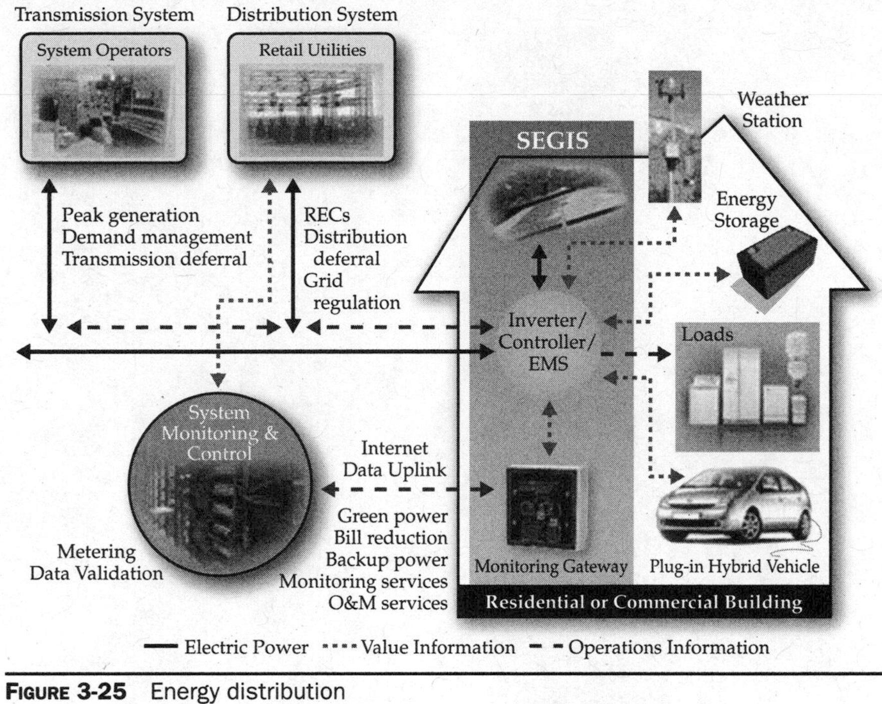

FIGURE 3-25 Energy distribution
http://www1.eere.energy.gov/solar/images/segis.jpg

CHAPTER 4

Energy-Efficient Home Systems

Identifying and modifying the most costly appliances and systems in your home can save you money and energy. You can reduce your electricity consumption by evaluating each appliance in your home and replacing the old energy hogs with new energy-efficient models. Think of your home as one complete system with components that affect how much energy you consume or save.

The chart shown in Figure 4-1 shows how much electricity is used each year by traditional home appliances. Replacing inefficient appliances with energy-efficient appliances, such as those that carry the Energy Star label and solar-powered sources, makes a lot of sense.

> Visit www.energysavers.gov for instructions on calculating the electrical use of your appliances.

Remove and Replace Appliances

The first stage in making your home more energy-efficient is to remove any unnecessary appliances—the old, second refrigerator in the basement, or the deep freeze, for example. Modern supermarkets are exemplary in providing fresh, suitably chilled food. Why deep-freeze beautiful fresh meats and vegetables until they are solid as a rock and as tasteless as cardboard? Why spend extra money to ruin perfectly good, fresh foods?

Every year, more energy-efficient appliances are available on the market. When purchasing a new appliance, you should always look for the Energy Star label. This does not guarantee that the appliance is the most energy-efficient appliance on the market, but an average Energy Star–

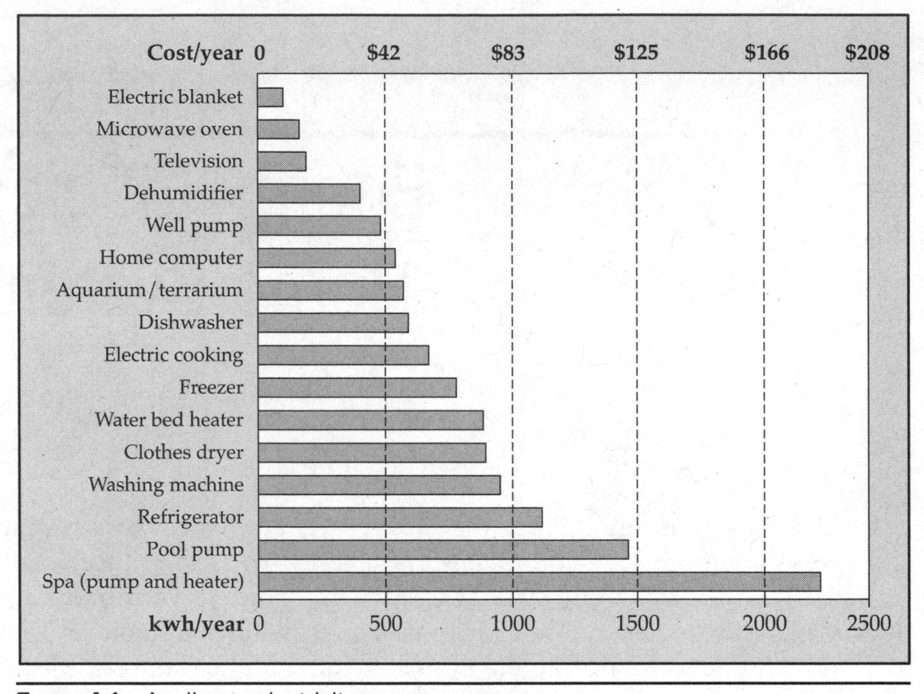

FIGURE 4-1 Appliance electricity use
Source: US Department of Energy, January 2009
http://www1.eere.energy.gov/consumer/tips/appliances.html

rated appliance uses approximately 40 percent less energy than the conventional models. For example, an Energy Star–rated refrigerator uses high-efficiency compressors, improved insulation, and more precise temperature and defrosts mechanisms to improve its energy efficiency.

The same is true for solar appliances. Although the efficiency of photovoltaic (PV) panels is not rapidly increasing, the appliances that use solar electricity are improving. In addition to the efficiency factor, many 12-volt applications including lighting, appliances, and charging stations are all more efficient than conventional applications. The efficiency is twofold: the electricity is produced onsite and with no loss of energy in transmission. Direct 12-volt usage means that electricity does not need to be converted from 120V AC to 12V DC, so no voltage loss occurs due to conversion. Any time energy is transferred or converted, some of the energy is lost in the conversion.

When replacing and disposing of an old appliance, take it to a local recycler for proper disposal or donate it to a reseller who refurbishes old appliances.

Washing Machines

Using energy-efficient washers can save you substantial amounts of energy, water, and money (Figure 4-2). Contemporary Energy Star–rated clothes washers are available in either front-loading or top-loading designs. Both types are highly efficient, but the most significant design difference for top-loading washers is that there is no central agitator.

Top-loading Energy Star–qualified washers use a system that flips or spins clothing through a stream of water. Front-loading washers tumble clothing through a small amount of water that resides in the bottom of the washer bin. Both designs use less water and therefore less energy. Advances in motor performance allow for more efficient wash and spin cycles. Faster spin designs extract more water and allow for less drying time.

FIGURE 4-2 Energy Star–approved washer and dryer
http://www1.eere.energy.gov/consumer/tips/images/pg26_washer_dryer.jpg

Clothes Dryers

Dryers are usually powered via gas or electricity, with gas dryers being more efficient than electric dryers. The best option is to choose an Energy Star–qualified model that is appropriate for your needs. Or you can choose a "solar-powered" drying system like the one shown in Figure 4-3.

FIGURE 4-3 "Solar-powered" clothes drying system
http://upload.wikimedia.org/wikipedia/commons/2/23/Clothes_line_with_pegs_nearby.jpg

Many of us remember childhood days when our parents hung laundry outdoors on the clothesline. Today, clotheslines have been replaced with "outdoor clothes dryers." A retractable clothesline is a transparent, convenient, low-cost option (Figure 4-4).

These clotheslines are sturdy and rust-free, and they can dry many loads at one time, which is not true for any indoor dryer. For a $50 investment, you can save hundreds of dollars each year when compared to using a traditional dryer.

Small Appliances

Many small electrical appliances used in the home can also be energy hogs. These appliances are usually purchased to provide a more pleasant environment in the home, or to help out in the kitchen:

- Dishwasher

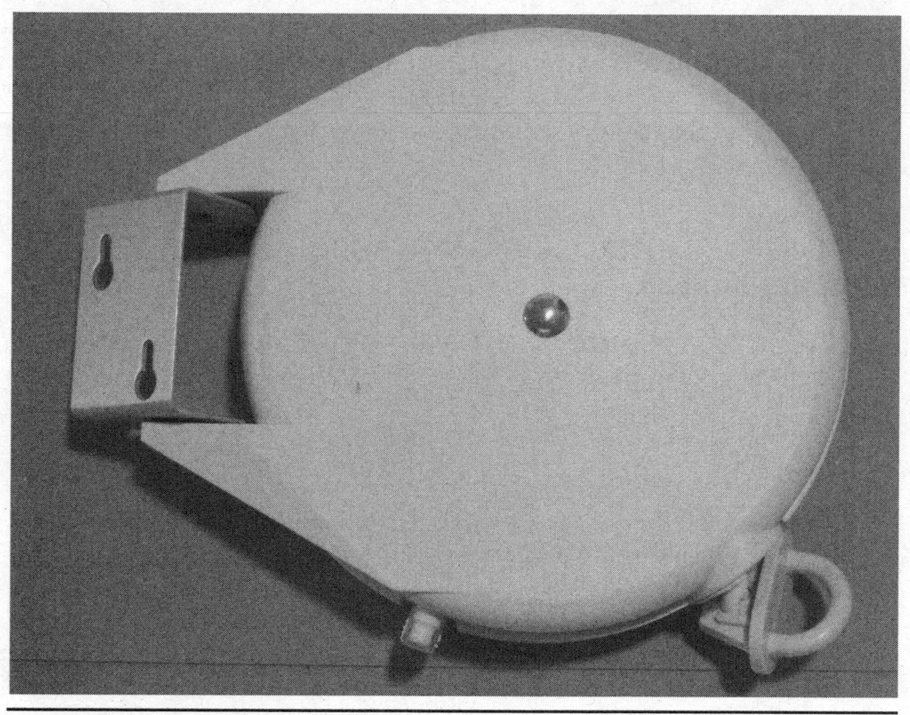

FIGURE 4-4 Outdoor dryer

- Room air conditioner
- Dehumidifier
- Room air filter
- Ceiling fan
- Microwave
- Toaster over

To save energy, consider the following recommendations:

- Buy appropriately sized appliances: too large and powerful, and you can overheat, overcool, or burn your food and waste energy.
- Always look for the Energy Star label. These appliances are usually made to higher specifications and use less energy. The energy savings of an Energy Star product will usually pay for itself over the lifetime of the appliance.
- Buy appliances according to your perceived needs or requirements. Understand how the appliance will be used, how often, and for what purpose. Is the new appliance necessary? This selection process will assist in the purchasing of the appropriate appliance.

In addition to their Energy Star rating, many small appliances can be purchased in 12-volt models. If you are installing a PV system in your home, you can take advantage of 12-volt appliances. Doing so will save energy production costs, transmission costs and losses, and conversion losses of energy.

Hot Water Heaters

Today you can choose from among many different types of energy-efficient hot water heaters:

- Solar water heaters
- Tankless water heaters
- Heat pump
- Gas condensing heaters
- Gas storage water heaters

Even though may other options are available, most Americans still use conventional water heaters (Figure 4-5) that are barely more efficient

FIGURE 4-5 Traditional water heater
http://www.cpsc.gov/cpscpub/prerel/prhtml03/03156a.jpg

than units sold 20 years ago. The average conventional water heater lasts approximately 15 years.

Instead of purchasing a conventional water heater, consider purchasing a solar water heater—a better option. The most advanced solar water heaters require little or no energy. The water is heated by the sun and stored in an insulated container similar to a conventional heater. Solar heaters can be either passive (Figure 4-6) or active collectors (Figure 4-7).

Solar Water Heater

Solar hot water collectors (Figure 4-8) are visible roof-mounted units that look similar to other types of solar panels. Instead of capturing the sunlight to produce electricity, the heat of the sun is captured and used to heat water or another solution. Solar hot water systems are designed for simplicity and efficiency.

Solar hot water systems use the sun's energy to heat water directly or indirectly via solar panels mounted in a position that will allow direct, long-term sun exposure. The sun heats antifreeze in the panels that then heats water via a heat exchanger. A heat exchanger allows the two different liquids to transfer heat but does not allow the fluids to commingle.

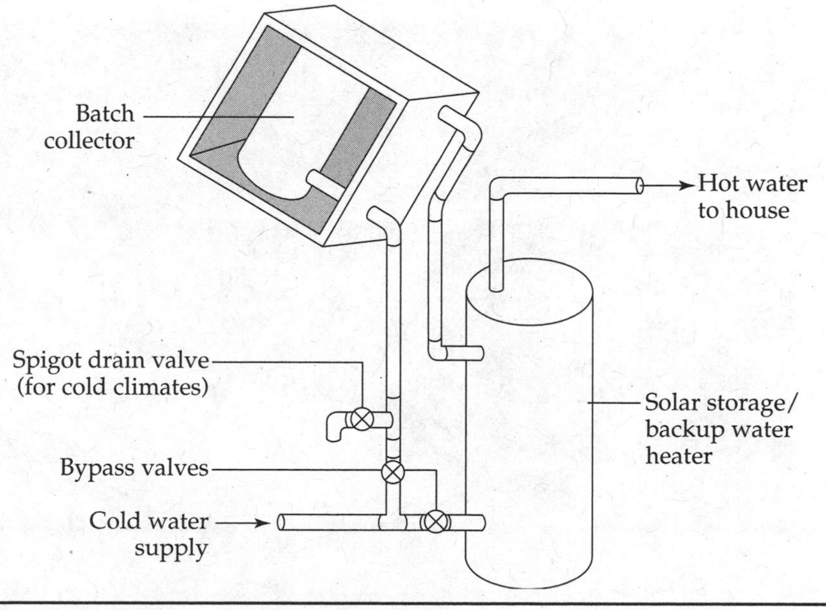

FIGURE 4-6 Passive, batch solar water heater
http://www.energyeducation.tx.gov/renewables/section_3/topics/solar_water_heaters/img/fig19p_solar_water.gif

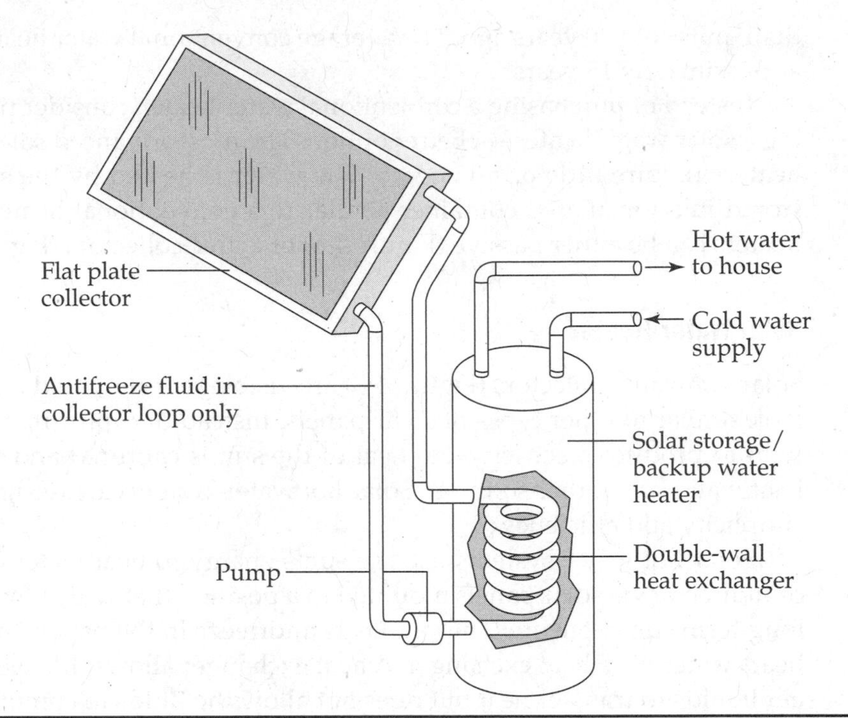

FIGURE 4-7 Active, closed-loop solar water heater
http://www.energyeducation.tx.gov/renewables/section_3/topics/solar_water_heaters/img/fig20a_solar_water.gif

FIGURE 4-8 Solar water heating panels
http://www.tucsonaz.gov/energy/IMG_0092.jpg

Active solar hot water heaters use electricity to power pumps to circulate the fluids through the system. Active systems are differentiated in three ways: a direct system that uses pumps to circulate water through collectors, an indirect system that pumps antifreeze through the solar collectors, or a drain-back system that heats the water and allows the water to drain into the holding tank. The active system provides the best comfort in mild climates, and the indirect and drain-back systems work well in colder climates.

Installing this type of product requires knowledgeable and experienced professionals. In addition, to qualify for tax rebates, you may be required to hire a certified solar installer to do the work.

When evaluating whether a solar water heater will work in your home, consider the following:

- Is direct sunlight available to the area of installation all year round?
- What type of system do you need, and what capacity?
- Does this type of system fit your lifestyle?
- What kinds of tax rebates or credits are available for each type of system?
- What are the maintenance requirements of each system?
- What is the final cost?

Tankless Water Heater

Tankless water heaters (Figure 4-9) are another popular alternative to traditional water heaters.

Tankless water heaters differ from traditional water heaters in that they do not contain a reservoir of heated water. Tankless units use gas or electricity to heat the water on demand. This means that the energy is used only when you require hot water. Traditional hot water heaters store large amounts of hot water. This is energy wasteful, however, because the hot water is not being used most of the time.

Because no tank is required, tankless units are great when space is limited. They are designed to function for 20 years or more. Tankless heaters may also quality for tax rebates, particularly if you install an Energy Star–rated system.

Two kinds of tankless systems are available: whole house and individual heaters. Similar to a conventional water heater, a whole house water heater is a single unit that is placed close to where you will use the water. The second option calls for multiple small tankless units installed at each location of use.

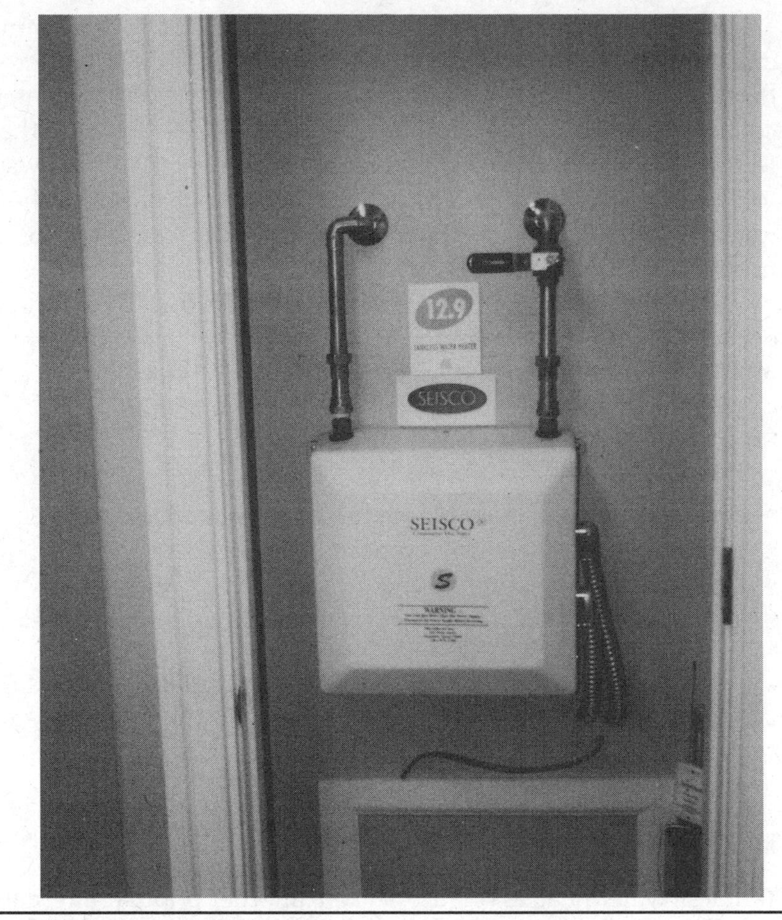

FIGURE 4-9 Tankless water heater
http://resourcecenter.pnl.gov/cocoon/morf/ResourceCenter/dbimages/full/592.jpg

Electric tankless water heaters are also available in 12-volt models that can be powered directly from a PV system.

Heat Pump

Heat pump hot water systems (Figure 4-10) offer another efficient alternative to traditional hot water heaters.

A heat pump system depends on a differential in temperature: A low-pressure liquid refrigerant is vaporized in the heat pump's evaporator and passed into the compressor. The refrigerant pressure and temperature increase. The heated refrigerant goes through a condenser in the storage tank that transfers the heat in the refrigerant to the water stored within the unit. When the refrigerant delivers its heat to the water, it cools

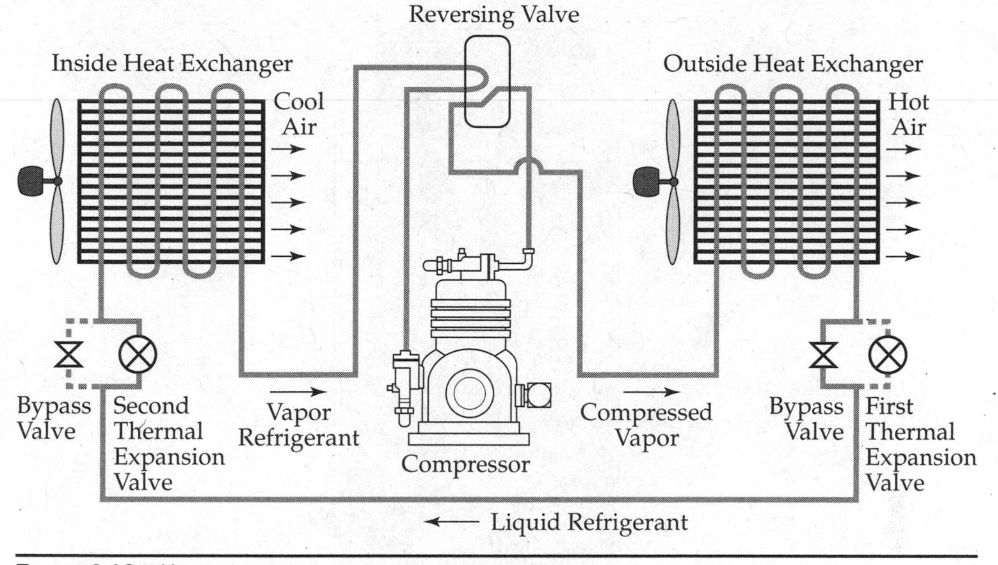

FIGURE 4-10 Heat pump
http://www.cdc.gov/nceh/publications/books/housing/Graphics/chapter_12/Figure12.01.jpg

and condenses. The refrigerant then passes through an expansion valve, where the pressure is reduced and the cycle begins again.

Electric heat pumps can be an excellent complement to PV solar power. They are often used in temperate climates, which provide ample sunlight for PV and moderate, continuous usage of electricity. The electricity is created and used onsite—the perfect complement of production and usage.

Gas Condensing Hot Water Heater

The gas condensing hot water heater (Figure 4-11) is similar to a conventional hot water heater with one exception: A conventional hot water heater vents hot gases as it heats water, but the gas condensing heater does not vent hot gases, instead reusing these gases to add heat energy to the stored water. The gas condensing hot water heaters are better and more efficient than traditional hot water heaters, but they are not as efficient as solar, tankless, or heat pump systems.

High-Efficiency Gas Storage Water Heater

High-efficiency gas storage water heaters are similar to traditional gas water heaters but have a few efficiency improvements. In these units, nat-

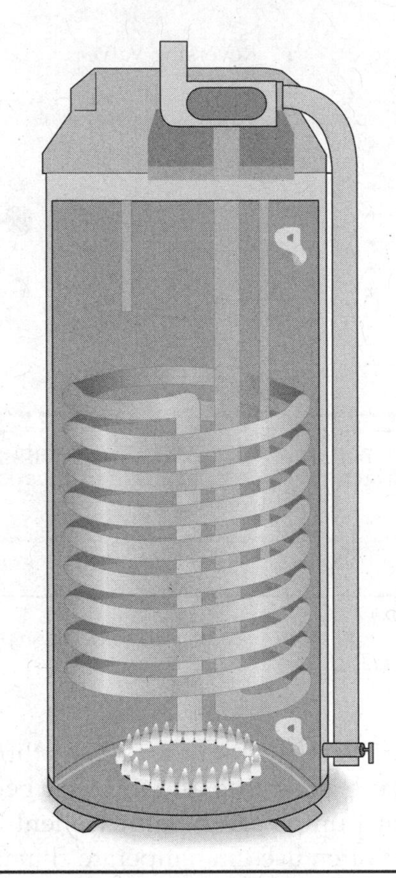

FIGURE 4-11 Gas condensing water heater
http://www.energystar.gov/ia/products/water_heat/images/GasCondensing_Works.jpg

ural gas is used to heat and store hot water for future use. Choose an Energy Star–rated system.

Cooling Systems

When it comes to cooling your home, two options are available: room air conditioners or central air conditioning.

Room Air Conditioners

When choosing a room air conditioner (Figure 4-12), the least efficient of the two choices, consider a few factors. Be sure that you choose a unit that will meet your needs: consider the size of the room(s) you want to cool and how well the room(s) is insulated.

FIGURE 4-12 Window air conditioner
http://www.cpsc.gov/cpscpub/prerel/prhtml01/01116.jpg

Air conditioners are rated in British thermal units (Btus). Most units display the following information on the box: the Btu rating, recommended square feet of coverage, and/or number of rooms that can be cooled. Consider the following as a rule of thumb:

- A 5000 Btu unit will cool one room.
- An 8000 Bty unit will cool three rooms.
- A 10,000 Btu or larger unit will cool an entire floor of a house.

When buying the unit, notice whether it requires 110, 220, or 12 volts of electricity. Be sure that you have a dedicated electrical outlet that provides the correct amount of amperage: using the wrong amperage could cause a fire.

Some small units are available in 12-volt models. When installing a PV system, include the option of allowing for expansion of 12-volt products that can use the electricity produced onsite.

When choosing an air conditioning unit, always buy an Energy Star–rated model, and look for tax credits or rebates before you purchase. Air conditioning is a luxury; however, the U.S. government will provide tax credits and rebates to help you enjoy your luxury efficiently.

Central Air Conditioning

Central air conditioning is more controlled, dispersed, and efficient than single room units (Figure 4-13). Although many energy- and environmentally-conscious people eschew central air, some people require it for health reasons or because the heat and humidity in some areas are almost unbearable.

Note that I do not endorse the installation of central air to be used continuously in a home. I recommend central air conditioning because in caparison to room air conditioners, central air conditioning is far more energy efficient. The proper-sized unit and appropriate use dictate the efficiency.

When purchasing central air conditioning, confirm the efficiency of the recommended units at the Energy Star website. When purchasing a new Energy Star–qualified central air conditioning system, purchase the appropriate-sized unit for your home. If you have an existing central air system and it is older than 12 years old, consider replacing it with a new Energy Star–qualified unit, which will pay for itself over its lifetime. Energy Star–qualified central air conditioning units typically use 30 percent less energy than other units.

FIGURE 4-13 Central air conditioning

Checking Ductwork in Your Home

An important part of a home's heating and cooling systems is the ductwork that runs throughout the home, as shown here. If you are replacing an older central air system, I recommend installing new energy-efficient, insulated ducts. Even if you believe your home's old ductwork is in excellent condition, have the system cleaned and checked; if necessary, have a contractor add more insulation to the system. This is also true if the cooling system shares the duct system with your furnace.

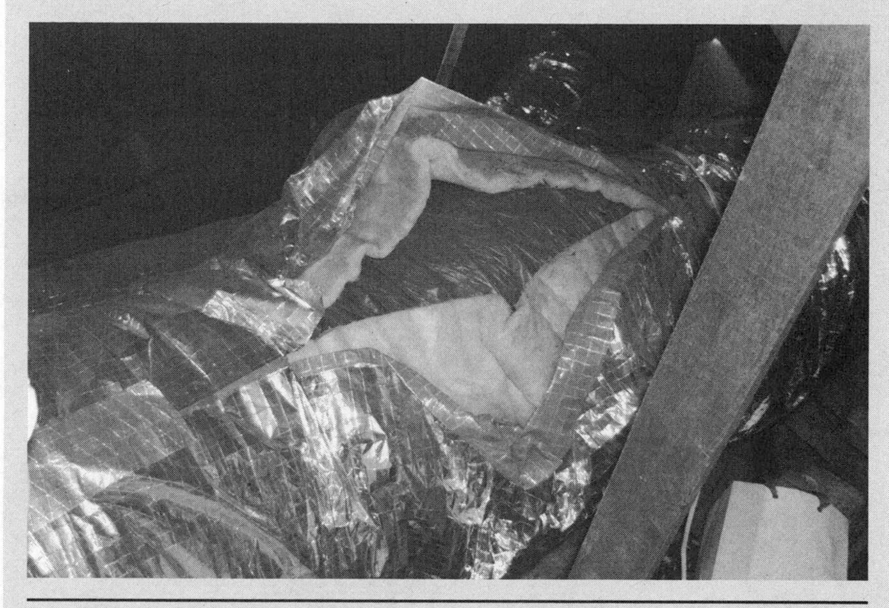

Central air conditioning soft duct

Sealing the duct system may seem easy, but this is not necessarily the case. Any time ducts are moved, the entire system can be disturbed. Moving or repairing one length can move a second length of ducting, causing fragile seals to separate from the area being repaired. Leaks can occur in one area after repairs are completed in a separate area.

Before installing or repairing any ductwork, call a contractor to check for leakage in the system. Insulation will not stop leaks. Air leakage, and the pressure differential caused by the leak, will deteriorate the duct insulation and create airborne particles that are hazardous to your health. A qualified contractor can validate that your duct system is leak-free by checking for leakage using a pressure differential test.

Sealing and insulating the heating ducts can increase the furnace's energy performance by as much as 25 percent.

For more information about duct sealing and product recommendation, the Energy Star website offers an excellent brochure at http://www.energystar.gov/ia/products/heat_cool/ducts/DuctSealingBrochure04.pdf.

Maintaining heating and cooling systems is an important task. You can lose all the efficiency gains if your system is not maintained. Have a professional examine your heating and cooling system at the beginning of each season and check the following:

- The refrigerant level
- The filters (Figures 4-14 and 4-15)
- The ducts (Figures 4-16 and 4-17)

FIGURE 4-14 Central air conditioning inlet and filter

The average home uses as much as half of its energy in the form of heating and cooling. Most of this energy is in the form of electricity. When choosing a heating or cooling system, or maintaining your current system, the average homeowner needs to be aware of many factors. Consumers incorrectly perceive that the heating or cooling unit is a single entity. The heating or cooling unit is only one part of the total home energy system that includes large appliances, all of which can be incorporated into a PV energy system for the most energy-efficient alternative.

FIGURE 4-15 High-efficiency particulate air filter HEPA

FIGURE 4-16 AC inlet duct

FIGURE 4-17 Soft duct in attic

Choosing the proper central air conditioning system for your home, using quality insulation, and scheduling regular maintenance will allow for years of energy-efficient and trouble-free comfort.

Heating Systems

Common residential heating systems come in the following forms: a furnace system, a boiler system, or a heat exchanger.

A furnace heats air using gas or electricity and delivers the hot air via air ducts to each room. The furnace is the most common heating system used in the United States. Energy Star–qualified units top the list of the most efficient furnaces.

A boiler heats water using gas or electricity and then distributes the heated water to radiators throughout your home. Boilers are common and efficient heating units, with those meeting the Energy Star requirements being the most efficient types.

Heat exchangers are often used in temperate climates and come in two forms: electric air source heat pumps and geothermal heating systems (Figure 4-18). Electric air source heat pumps are often used in mod-

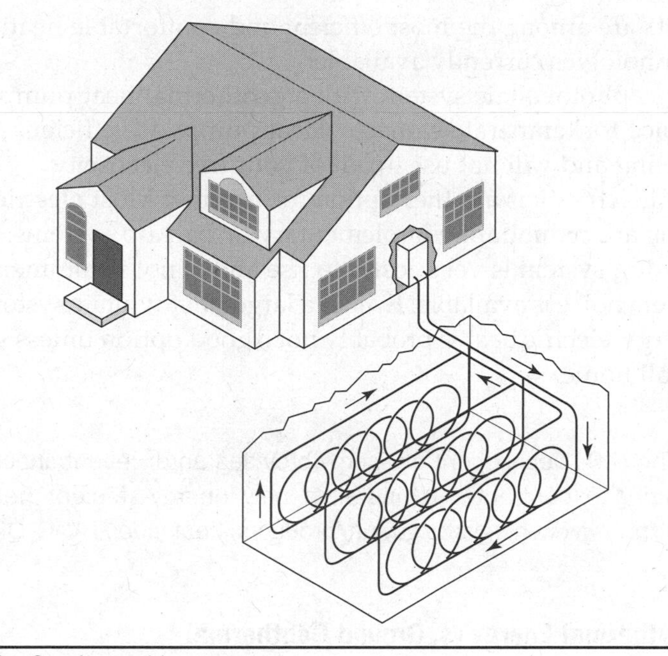

FIGURE 4-18 Geothermal heat system
http://www1.eere.energy.gov/buildings/residential/images/geothermal_energy_1.gif

erate climates; these units use the difference between outdoor air temperatures and indoor air temperatures to cool or heat your home.

Heat exchangers are efficient ways to heat and cool a home. The immediate benefit is that you need only one system to perform both functions. The limitation of these systems is climate, because they rely on a differential in temperature. Much like a Stirling engine, you need both hot and cool areas for these units to produce the required results.

What type of heating system is correct for you? For small, individual residential homes, small complexes, and small apartment units, a furnace is probably the most effective choice. A furnace should be a central heating unit with short duct systems that allow for the efficient transfer of heat to each room of the home, complex, or unit.

Boilers are the next best choice for heat. If heat must be transferred long distances—such as to units in an apartment building—a boiler is the best choice. Boilers can also be used to supply hot water. A furnace system requires a separate hot water heater.

Geothermal heat pumps use heat generated from the ground to heat and distribute heated water or air. These units are commonly used to distribute hot water, but they can also be used to distribute hot and cool air. Because geothermal units use the earth's natural stable temperature, these units are among the most efficient and comfortable heating and cooling technologies currently available.

A photovoltaic system with a geothermal heat pump is an excellent choice for temperate climates. Heat pumps are efficient for heating and cooling and will not use up all of your free electricity.

Electricity is another option for heating. Most electrical heating systems are redundant complementary or backup systems. Electricity as a heating system is very costly to use and is not recommended, unless no other choice is available. Even if a large photovoltaic system produces the energy, electric heat is probably not a good option unless you have a very small home.

The U.S. Department of Energy provides an excellent checklist regarding hiring a contractor to install a new energy-efficient heating system: http://www.energystar.gov/ia/products/heat_cool/HVAC_QI_bidsheet.pdf.

Geothermal Energy vs. Ground Geothermal

Before I explain how geothermal heating and cooling actually work, you need to understand the difference between geothermal energy and ground geothermal. Geothermal energy is produced by drilling large wells deep into the ground and using the hot water and steam created there to generate electricity (Figure 4-19).

FIGURE 4-19 Geothermal power station
http://www.nrel.gov/data/pix/Jpegs/00427.jpg

The spectacular geyser in Yellowstone Park, "Old Faithful," is an excellent example of geothermal underground heat. Most geothermal heat is located deep within the Earth near fault zones. This type of energy is completely free except for the development costs, produces zero emissions, and is a completely carbon-free energy source. The problem is that it is a economically viable solution in only a few areas of the world.

Ground geothermal, on the other hand, is a heating and cooling source that is available to most consumers (Figure 4-20).

Ground geothermal energy depends on a temperature differential—the inside of your home must be at a temperature that differs significantly from the ground temperatures outside your home (which is usually around 60°F). During the summer, the ground source heat pump can actually cool your home.

The gas in the heat pump is compressed into the pipes outside of the home until it reaches approximately 130°F. Because the ground surrounding these pipes is at 60°F, and the pipes are in direct contact with the ground, the heat is removed through conduction. Eventually the coolant/gas will chill to ground temperature. The gas is then allowed to expand through the pipes inside the home. When a gas expands it cools, releasing cool energy. The process is repeated until the home reaches the appropriate temperature.

In the winter, the system is reversed. The gas in the geothermal system is compressed inside the home. When the gas is compressed, it rises in temperature, heating the home. This system is efficient because it re-

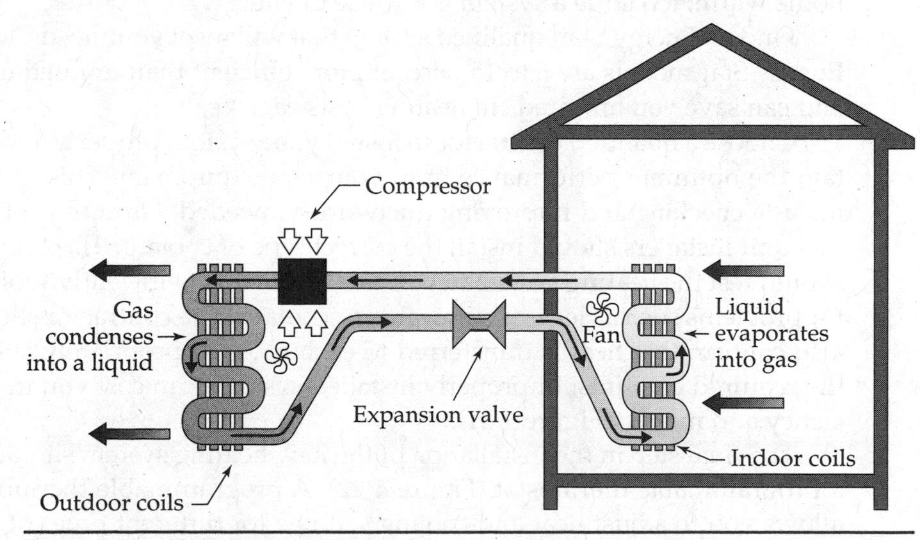

FIGURE 4-20 Heat pump cooling cycle
http://www.energysavers.gov/images/split_system_heat_pump_cooling.gif

quires only a small compressor to exchange the same amount of gas in reversing directions.

Maintaining Home Heating and Cooling Systems

For all home heating methods, you need to follow some simple rules to make sure the systems are operating at maximum efficiency. Poor performance or premature failure of your heating system can occur because of a lack of maintenance.

Yearly maintenance, or a tune-up, for the heating system is required. You or a professional should inspect the main unit as well as the inlet and delivery systems. A poor performing or dirty furnace or boiler can use 20 percent more energy than it should if it were performing well. Change any filters at regular or required intervals.

> You can find an excellent checklist for servicing your heating needs at: http://www.energystar.gov/index.cfm?c=heat_cool.pr_maintenance.

What to Look For in a New Heating System

If you are ready to replace your heating system, first choose the type of system that best meets your requirements. Talk to a professional to determine what size system is best for your home. If a new heating system is too small to support the size of your home, it won't be able to keep your home warm; too large a system is a waste of energy.

Find an Energy Star–qualified system that will meet your needs. Most Energy Star models are 6 to 15 percent more efficient than unrated units and can save you hundreds in heating costs each year.

Choose a qualified contractor to install your system properly and obtain the optimum performance from your new equipment. This should include checking and improving ductwork as needed. Ground geothermal unit installers should install the correct type of coolant. The installer should test the heating system to be sure that it works efficiently, looking for problems such as leaks of air, water, or coolant. The contractor should also confirm that heat is transferred to each room appropriately and in the required amounts. Improperly installed systems can cost you in efficiency and money (Figure 4-21).

The final step in the installation of the new heating system should be a programmable thermostat (Figure 4-22). A programmable thermostat allows you to adjust heat and cooling settings for different times of day and night to make the most of your system.

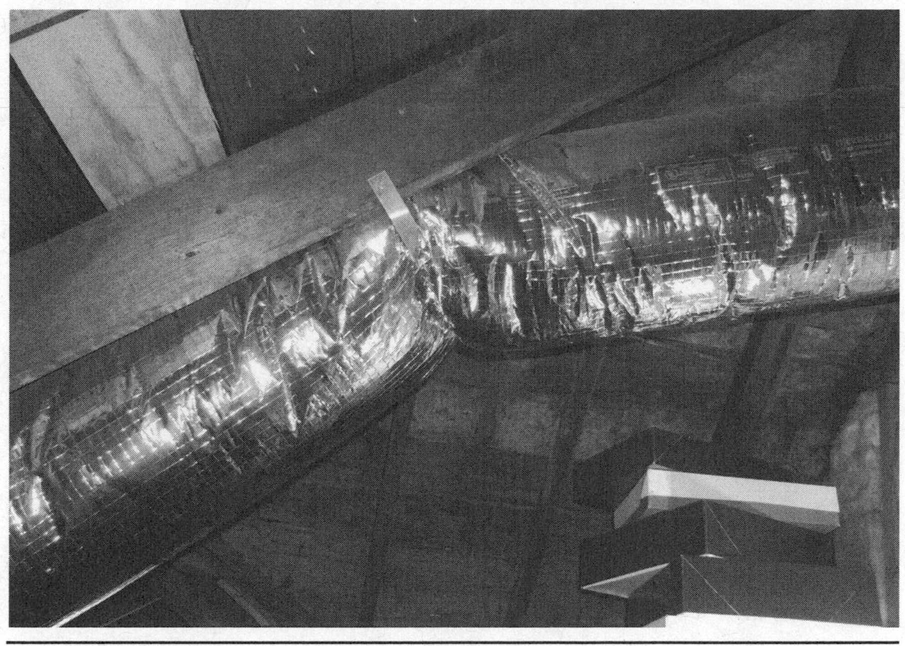

Figure 4-21 Ductwork
http://www.energystar.gov/index.cfm?c=heat_cool.pr_hvac

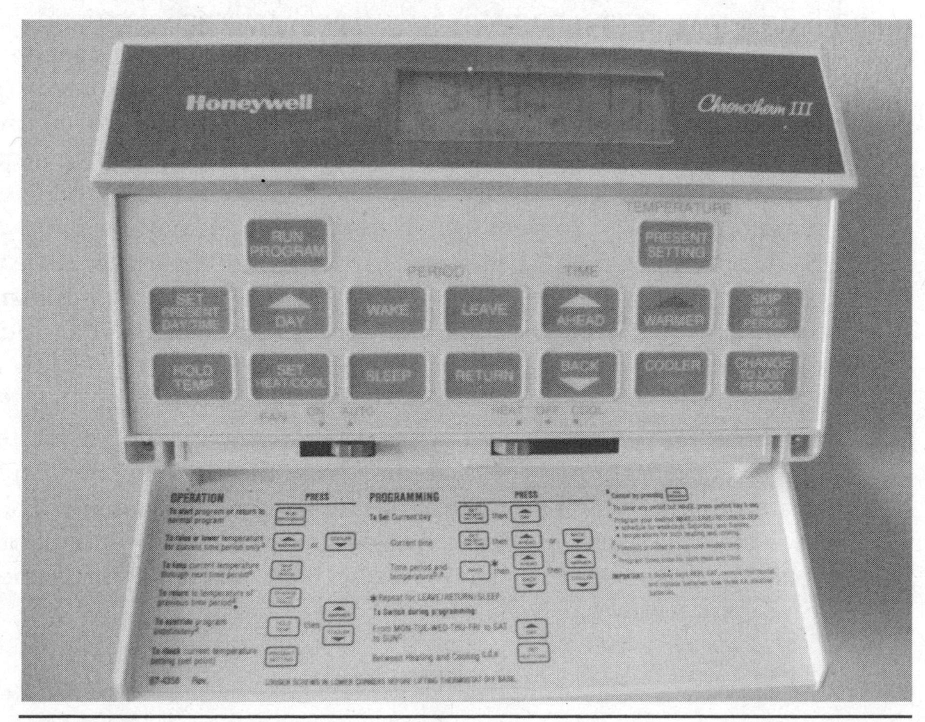

Figure 4-22 Programmable thermostat

A programmable thermostat can be integrated into your PV system as well and can be set up by your PV installer.

Other Large Energy Consumers

A few large energy devices are not typical of every household, but they deserve mention because they consume large amounts of power. These items are not required for basic home use but many consumers buy these "luxury" items for their homes:

- pool pumps
- aquariums
- electric blankets
- tanning beds
- heat lamps
- exercise equipment
- fans

(Chapter 8 shows how you can choose to make some of these energy consumers dependent or independent of your current electrical system.)

These items consume more energy than the average electric appliance because of the amount of power required or the length of use. Pumps, motors, and compressors all require large amounts of power. A pool pump may require significant power but may also be required to run for 8 to 12 hours each day. An aquarium may require pumping and heating.

Many homes have exercise equipment, which has powerful energy-consuming motors. Depending on your habit, it may be more cost-effective to join a gym. Of course you can always go for a walk, since your taxes already pay for the street. Or you can use exercise equipment to power other appliances: hook up a stationary bike to pedal or a treadmill to walk to create power, and use that power to watch TV or operate a clothes dryer.

Hopefully, you now have a better understanding of how you use energy, or more specifically electricity, in your home. Reducing your energy consumption is the first choice in home improvement. Installing a PV system to meet the remaining needs is a good complement to your efforts. In the next chapter, you'll learn how you can increase your comfort and decrease your energy bills with zero-cost passive solar.

CHAPTER 5

Zero-Cost Passive Solar

Passive solar is the most simplistic and yet overlooked form of energy—not just solar energy, but all energy—even though passive solar energy is actually the most reliable form of energy.

In passive solar energy systems, the sun's energy is used without any mechanical systems or human intervention. Today, humans consume less "fresh" energy in the form of sunlight and more "stale" energy in the form of fossil fuels, than at any time in history. We have ignored the fuel that is abundant and free—sunlight—and have subsidized our own demise (Figure 5-1).

The solution is simple: we can open the blinds (Figure 5-2). We do not need to spend tens of thousands of dollars redesigning our homes, because that would waste energy. We need to take advantage of what we have and improve our homes in renewable, energy-efficient ways. Replace and buy new when items are severely inefficient or in need of replacement from normal use. And when you buy, do so wisely.

You can design your home with solar energy in mind or make changes in your home to make the most of this free source of energy. This chapter discusses some great passive solar ideas about what you might be able to do in your own home. You will be surprised at the modest changes you can make that can dramatically impact your home, your comfort, and your pocketbook.

Figure 5-1 Life energy
http://apod.nasa.gov/apod/image/0110/anticrepuscular_britton_big.jpg

Figure 5-2 Natural sunlight
http://www.arts.gov/bigreadblog/wp-content/themes/default/images/Camden.jpg

Passive Solar Homes

Passive solar homes are nothing new. The home shown in Figure 5-3 does not offer any of today's modern amenities, but it does demonstrate two great features that you can use today in your passive solar home. This ancient home was built into the side of a cliff, and the wall inside the cave was built by humans. The wall shelters the occupants from the harsh sun of the day, while letting moderate sunlight into the home. The same wall

Figure 5-3 Ancient passive solar home
http://www.nature.nps.gov/GEOLOGY/parks/gicl/images/gicl_dwell.jpg

collects heat during the day and then releases that heat in the evening. This simple feature is used extensively in passive solar today, as it was thousands of years ago.

Today we use bricks, concrete, glass, and wood to build our homes (Figure 5-4). The sun heats these materials during the day and the stored heat is released in the home in the evenings. We use building materials, as our ancestors did, to control the environments in our homes.

Figure 5-4 Passive solar today
http://www.oregon.gov/ENERGY/RENEW/Solar/images/SpaceThumb.jpg

The Big and Small Pictures

On a global scale, passive solar is a part of the global warming problem (Figure 5-5). Heat energy enters the atmosphere, heats the Earth, and then radiates back into the atmosphere. Unfortunately, because of the layer of greenhouse gasses lingering above the Earth's surface, the heat has nowhere to escape. The trapped heat then increases the temperature of the Earth every day, as more thermal energy is supplied by the sun.

Passive solar heating, whether in a home or an actual greenhouse (Figure 5-6), works in a similar way. In a greenhouse, passive solar light and heat are allowed to enter the space. The quantity of heat and light that passes through is based on square feet of available sunlight and the angle of the sun. Both light and heat can be controlled, via shade or by venting or fan-cooling the greenhouse. Like a greenhouse, a home can be configured to capture sunlight (Figure 5-7) and its energy.

Building a Passive Solar Home

The first two considerations when you're planning to build a passive solar home are the climate and the solar altitude of the area where you want to build (Figure 5-8). A passive solar home can be built to accommodate average or severe climates. The solar altitude concerns angles of the

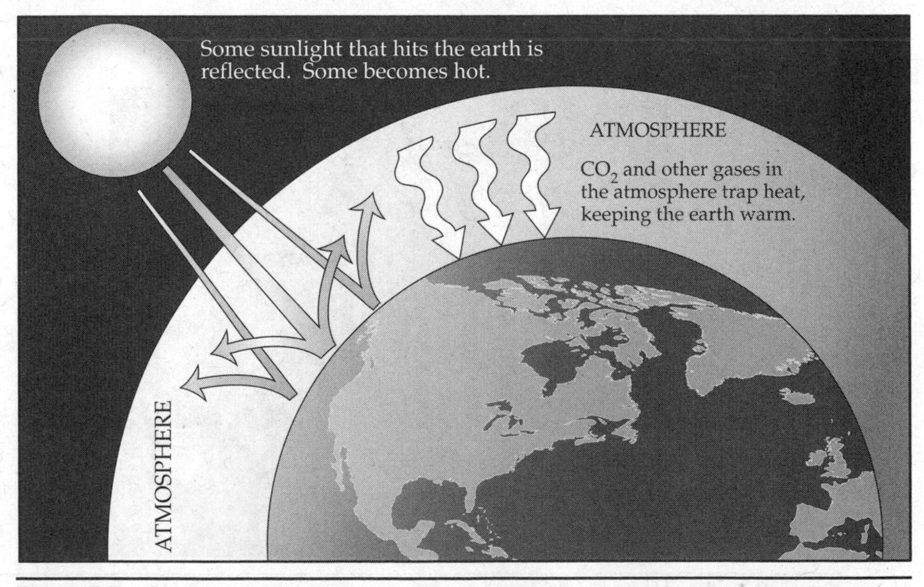

FIGURE 5-5 Global warming
http://www.ecy.wa.gov/climatechange/images/greenhouse_effect2.jpg

FIGURE 5-6 Greenhouse
http://www1.eere.energy.gov/tribalenergy/guide/images/photo_greenhouse_co.jpg

FIGURE 5-7 Passive solar in a home
http://www.coconino.az.gov/uploadedImages/Community_Development/Stephanie/Bippus%20Residence.bmp

sun, which will affect how much light and energy passively enters into the home.

Your next consideration is the orientation of the home, which is also related to the climate and the solar altitude (Figure 5-9). These factors must be addressed by your architect before the building design begins.

A passive solar home has five elements (Figure 5-10): a thermal mass to reflect light and collect heat during the day; an absorbing material to hold and store the heat; an aperture or opening to allow the light energy to enter the home; a control to vary the amount sunlight that is passed;

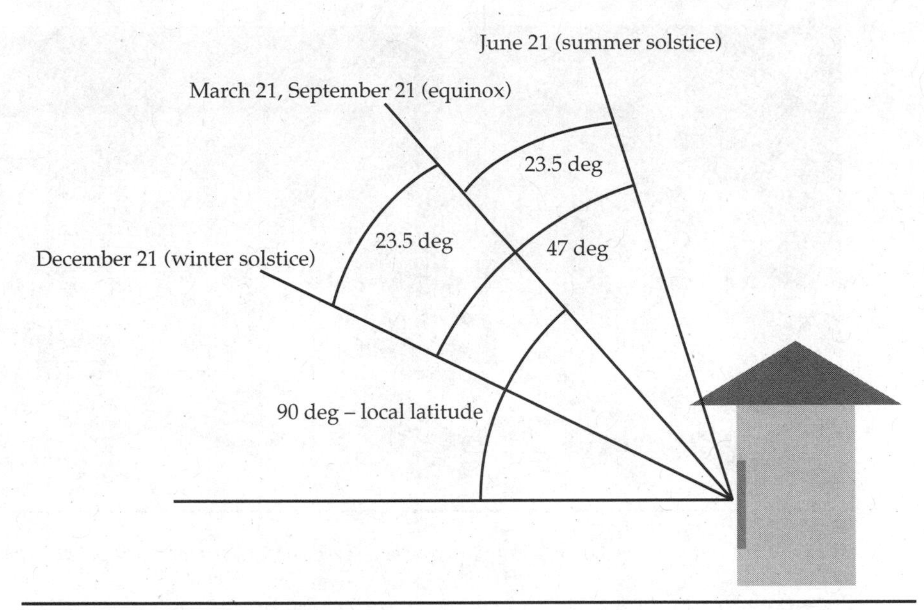

Figure 5-8 Solar altitude
http://en.wikipedia.org/wiki/File:Solar_altitude.svg

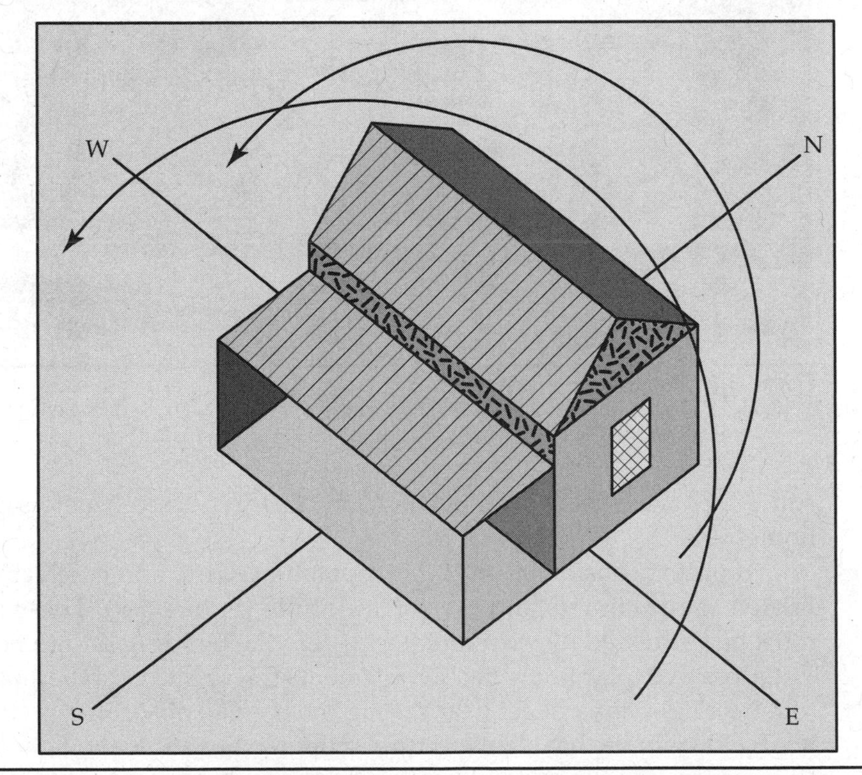

Figure 5-9 Passive solar home orientation
http://dnr.louisiana.gov/sec/execdiv/TECHASMT/ecep/comfort/c/com-c2.gif

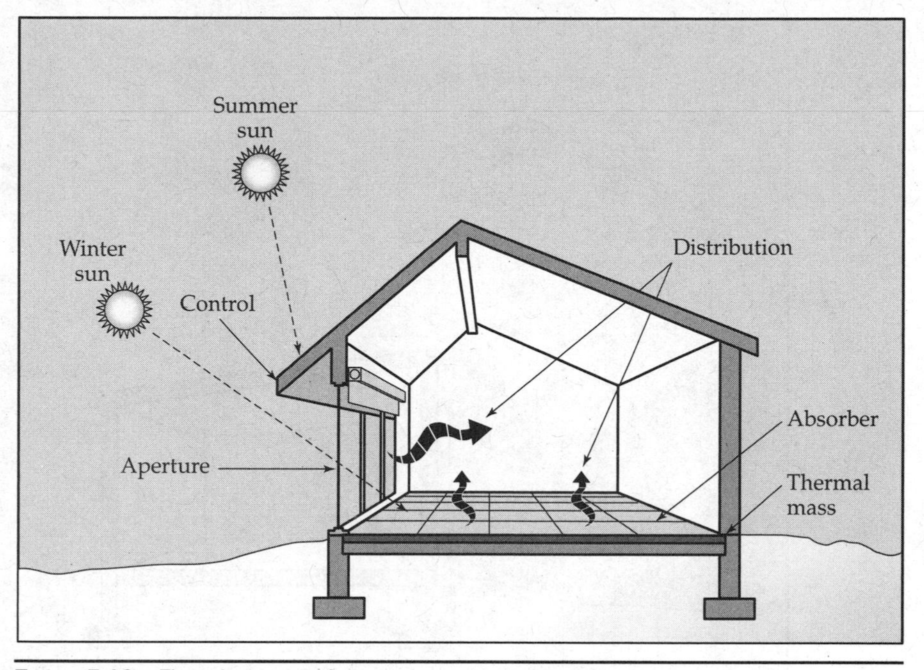

FIGURE 5-10 Five elements of a passive solar home
http://www.energysavers.gov/images/five_elements_passive.gif

and a method for distributing energy during the evening. These elements must be present and designed correctly for passive solar to function without intervention or mechanical assistance.

With proper home maintenance, passive solar elements will last indefinitely—no furnace to replace every 20 years, no electricity required, no motors to wear out, and no noise, smell, or toxic chemicals involved. Passive solar is clean, simple, efficient, and free.

The same process can be achieved in the vertical (Figure 5-11) using a Trombe wall, named after French engineer Félix Trombe, who popularized its use in the 1960s.

These are the basics of designing a passive solar home. You can then design the appearance of the home with your architect, the same as any other home.

The details in passive solar energy are in the design and proper implementation. Understanding the concept and knowing the five components of passive solar should help you understand and use passive solar in your home design.

If you would like assistance or more detail on passive solar design, you can e-mail me at Solar@exploresynergy.org.

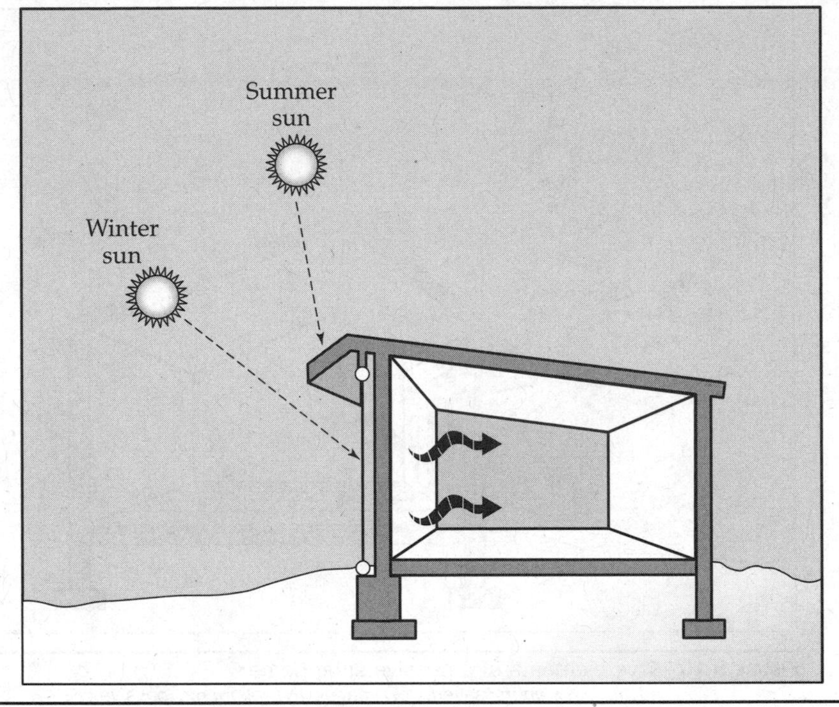

Figure 5-11 Vertical thermal mass
http://www.epa.gov/ne/eco/energy/images/trombe-wall.jpg

Passive Solar Home Modifications

There is a happy medium between designing and building a brand new home and merely opening the blinds to let the sun into your current home. You can build a passive solar addition to your residence, such as a sunroom (Figure 5-12)—a simple and effective way to harness the sun's energy. Many sunroom plans and kits can be purchased and installed.

Figure 5-13 shows all five of the components of passive solar in a single room addition. This type of home addition is a relatively low-cost option. The price of a new home may begin at $200,000, but a passive solar addition can cost only a few thousand dollars.

Another option is to modify an existing room in your home to make it more solar friendly. Your home is probably built in a square or rectangular format, and one or more sides of your home will likely have some sun exposure. Add a thermal mass floor or wall and some properly placed windows, and you can create a room that most people will envy—a warm, quiet, sunny place to sit during a cold winter day. This type of home modification can often be accomplished without disruption to the rest of the home, and at a relatively low cost.

FIGURE 5-12 Sunroom
http://resourcecenter.pnl.gov/cocoon/morf/ResourceCenter/dbimages/full/972.jpg

Good Solar Design

Light and thermal energy are gained in a solar home or room through radiation. The transfer of that energy takes the form of convection. A properly designed space must allow the correct amount of light in and control the amount of energy dissipated.

Insulation allows for the containment of energy taken in. Radiation absorption can be increased passively with color. The *albedo* of an object refers to how strongly it reflects light (Figure 5-14). For example, a white surface will reflect sunlight almost nine times more than a black surface. This means that your thermal mass, the area that stores heat energy, should be dark in color. A light-colored floor or wall may rise in temperature to 80 or 90°F, but a dark-color thermal mass can rise in temperature to 130 or 140°F.

Types of Solar Gain

In indirect solar gain, heat enters the building through an aperture and is captured and stored in a thermal mass. The mass then slowly and indi-

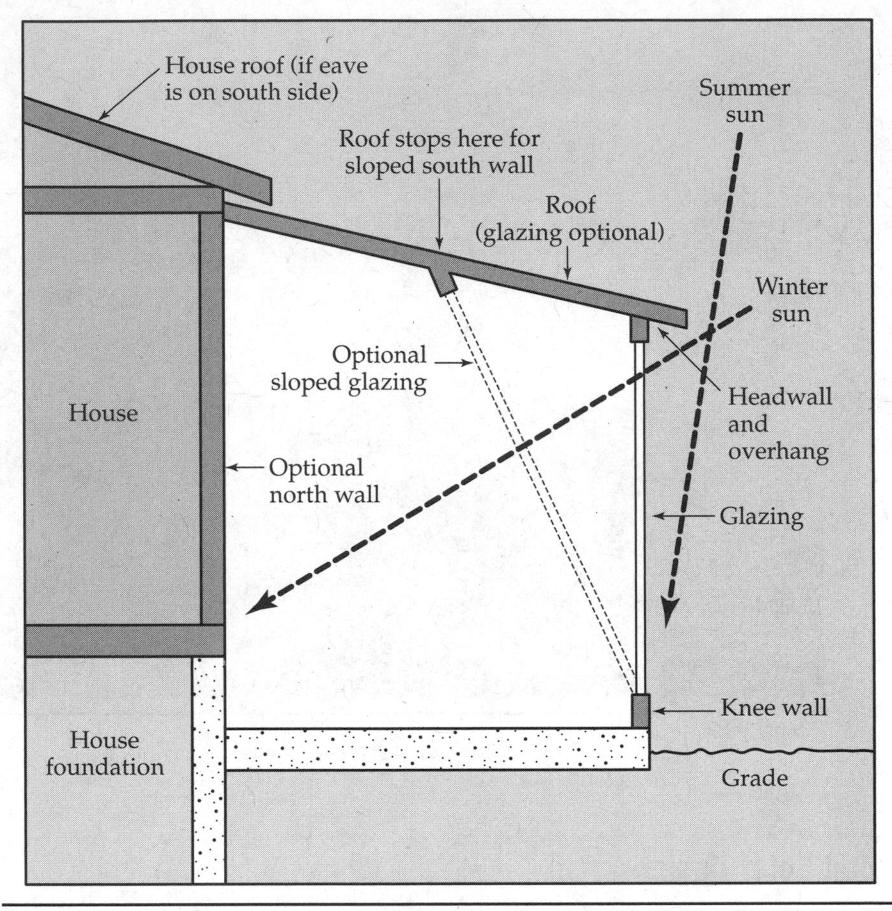

FIGURE 5-13 Passive solar home addition
http://www.energysavers.gov/images/overhang.gif

rectly heats the building through conduction and convection. Isolated solar gain is a separate space, such as a sunroom, in which the solar energy is captured and then passively moved as heat through the living space by natural convection.

A third type of solar gain, concentrated solar power (CSP), was reviewed in Chapter 3. While most applications are large commercial structures, consumers can use CSP water heaters in their homes. CSP is best in warm climates where sunlight is a constant and rainfall is low.

Simple Improvements for Solar Gain

Every homeowner and renter can make simple home improvements to increase the passive solar energy available and used in the home. Keep-

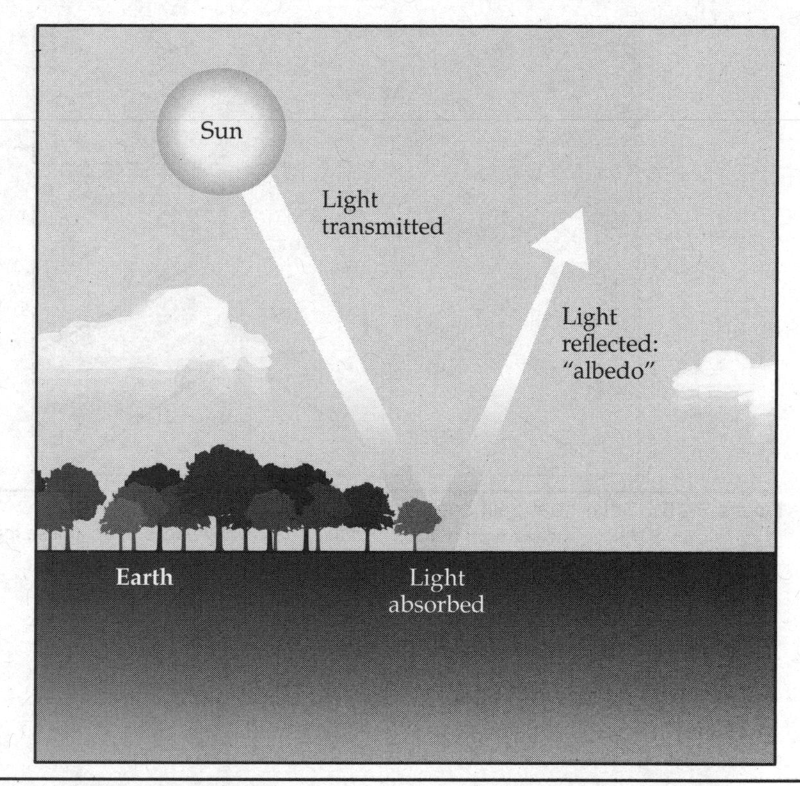

FIGURE 5-14 The albedo effect
http://www.energyeducation.tx.gov/environment/section_3/topics/predicting_change/img/albedo.gif

ing in mind the five elements required for passive solar, consider the following.

The thermal mass in most homes is already in place and will not change. This solar storage will be a floor or wall. The larger and more dense the mass, the more heat storage is available. You can also change the color to absorb more heat energy. For most homes, the aperture is in the form of windows or skylights, and the control will be some type of shade.

Heat and Light Transfer Through Windows

The solar heat gain coefficient (SHGC), shown in Figure 5-15, refers to the percentage of solar radiation that passes through a window and is expressed in a value of 0 or 1. The higher the SHGC, the more solar gain in the home.

The visible light transmission (VT), shown in Figure 5-16, refers to the amount of light transmitted through a window and is also measured by values between 0 and 1. A window with a VT of 1 will allow no loss of

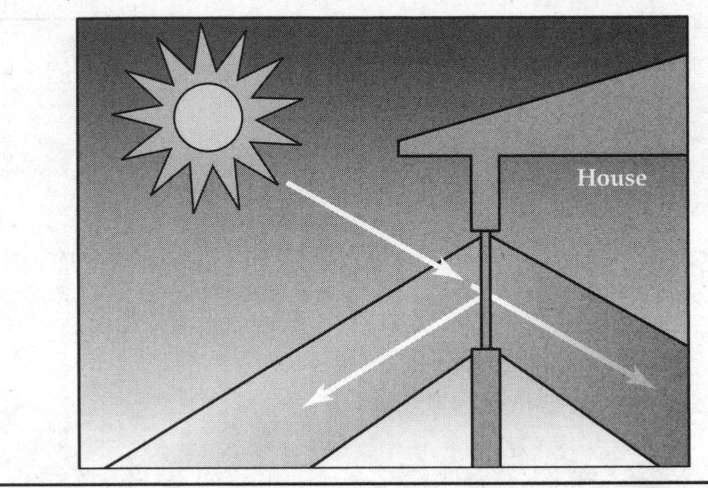

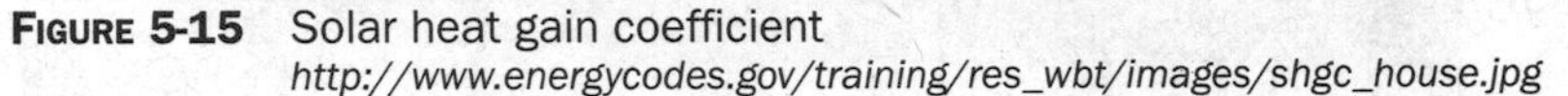

FIGURE 5-15 Solar heat gain coefficient
http://www.energycodes.gov/training/res_wbt/images/shgc_house.jpg

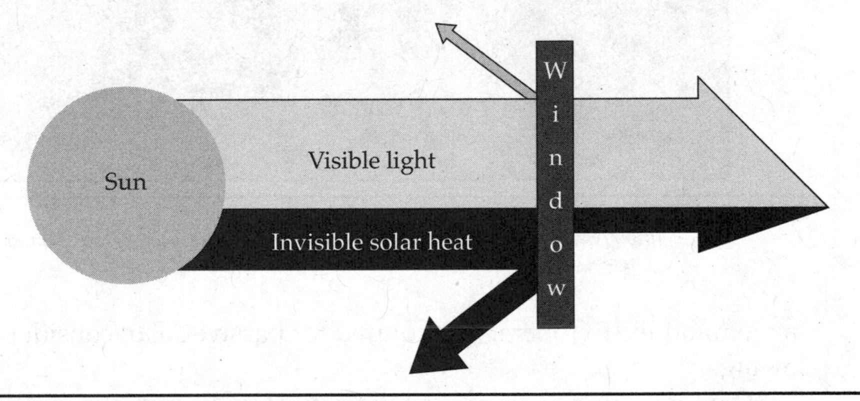

FIGURE 5-16 Visible light transmission
http://eetd.lbl.gov/lab2mkt/images/spectral-glazing-t.gif

transmission. The use of a second pane of glass or a coating will lower the VT value.

Low-E coatings, which stands for low emissivity, refers to the relative ability of a surface to emit energy by radiation. A Low-E coating for windows refers to a thin metallic or metallic oxide layer deposited on one of the glass surfaces (Figure 5-17). Heat gained or lost through a window is due to radiation from either outside-in or inside-out. The coating reflects a portion of the radiation, reducing heat flow through the window. The coating is normally invisible to the homeowner. Low-E is also rated on a scale of 0 to 1. A black object would have an emissivity of 1, absorbing all light, and a perfect reflector would have a value of 0.

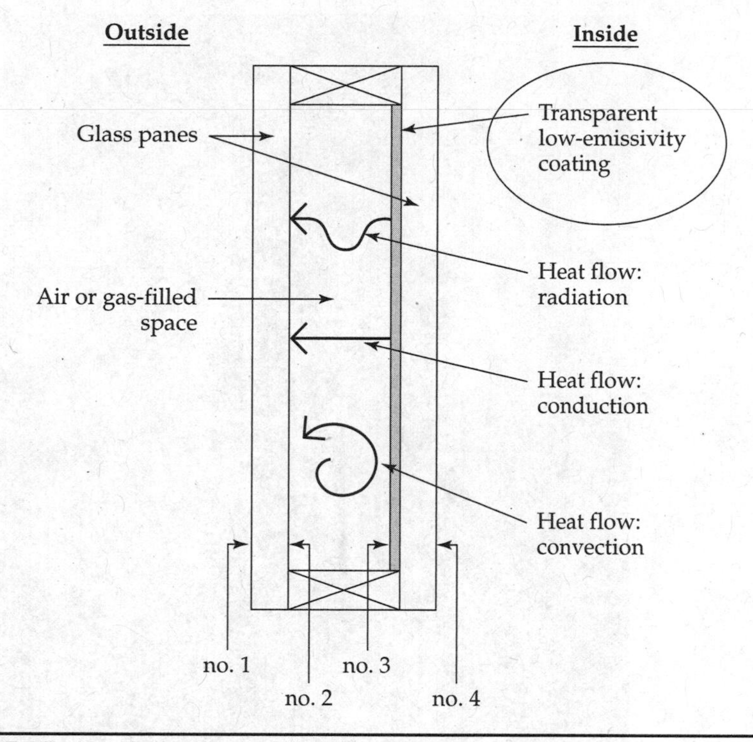

Figure 5-17 Low-E window coatings
http://pag.lbl.gov/images/Low-E%20window.jpg

A single-pane window is a single sheet of glass with no coatings (Figure 5-18). Single-pane windows are energy inefficient, but in the last 30 years, dual-pane and triple-pane windows have become popular due to their increased efficiency at restricting the transfer of heat.

The U-value gauges how well a window conducts heat and is more correctly called the overall heat transfer coefficient. It measures the rate of heat transfer through a window (Figure 5-19). A window's U-value is rated under standardized conditions. The usual standard is at a temperature gradient of 75°F, at 50-percent humidity, with no wind. The lower the U-value, the less heat transfer occurs, and the more energy-efficient the window is considered to be.

Here are examples of typical window performance:

- A single-pane window has an average U-value of 1.1 or greater.
- A dual-pane window has an average U-value of 0.55 or lower.
- A triple-pane window has an average U-value as low as 0.20.

In addition to the glass, the window frame, which is included in the U-value, must be designed for performance (Figure 5-20).

FIGURE 5-18 Single-pane window in older home
http://dpddata1.seattle.gov/dpd/Apps/HistoricPhotos/030-0608.jpg

A common window used in the 1960s was an aluminum-framed window with a single pane of clear glass, and this window is used as our standard for comparison for the following statistics, which approximate averages:

- A wood or vinyl frame with dual-pane clear glass would be 27 percent more efficient.
- Wood or vinyl frame with dual-pane glass and a Low-E coating would be 32 percent more efficient.
- An insulated frame window with triple-pane glass and a Low-E coating would be 41 percent more efficient.

Orientation can also have a great effect on how the light strikes the windows and heats or cools the house. In a cold climate, south-facing

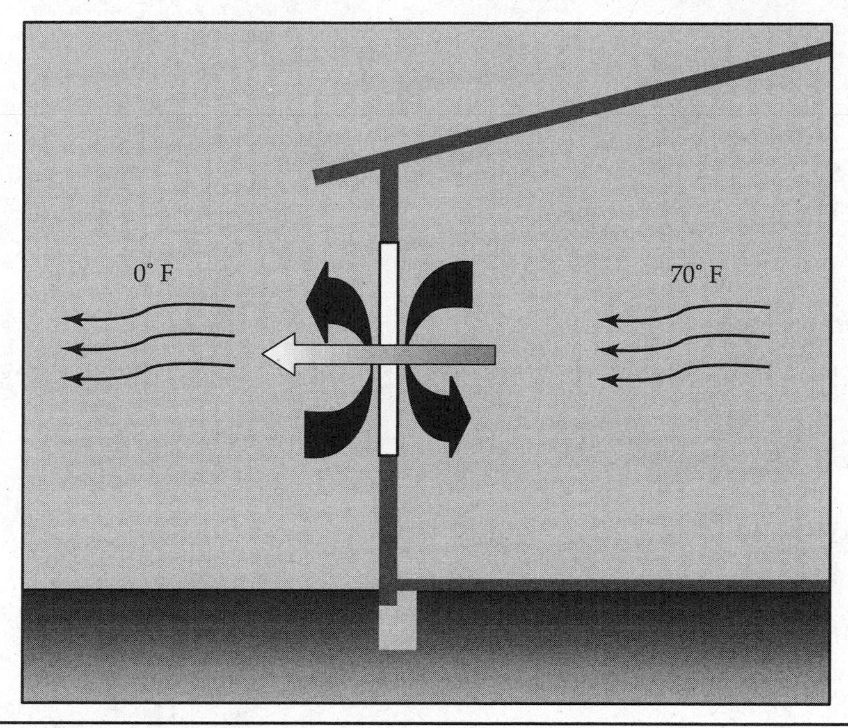

FIGURE 5-19 Heat transfer
http://resourcecenter.pnl.gov/cocoon/morf/ResourceCenter/dbimages/full/10.jpg

windows with a high SHGC can assist with winter heating costs provided you purchase windows that will allow for solar gain. In a warm climate, north-facing windows with a low SHGC are more suitable. East- and west-facing windows are more difficult to shade.

Gas-Filled Windows

In gas-filled windows, two or more panes of glass surround a sealed space that is normally filled with an inert or noble gas (Figures 5-21 and 5-22). Nitrogen is most common, followed by argon, and krypton, with their heat transfer properties from high to low, respectively.

Two layers of glass are the industry standard in warmer or cooler climates, while triple-pane sealed glass has the lowest heat loss characteristics and is more efficient in the hottest and coldest climates.

Nitrogen gas is normally used in the window fill space. Argon is also used and has a lower heat transfer characteristic than nitrogen. It slowly evaporates from the cavity over time, causing a loss of insulating characteristics, but it is generally accepted that until the argon concentration is less than 75 percent, usually in about 20 years, the window retains its abil-

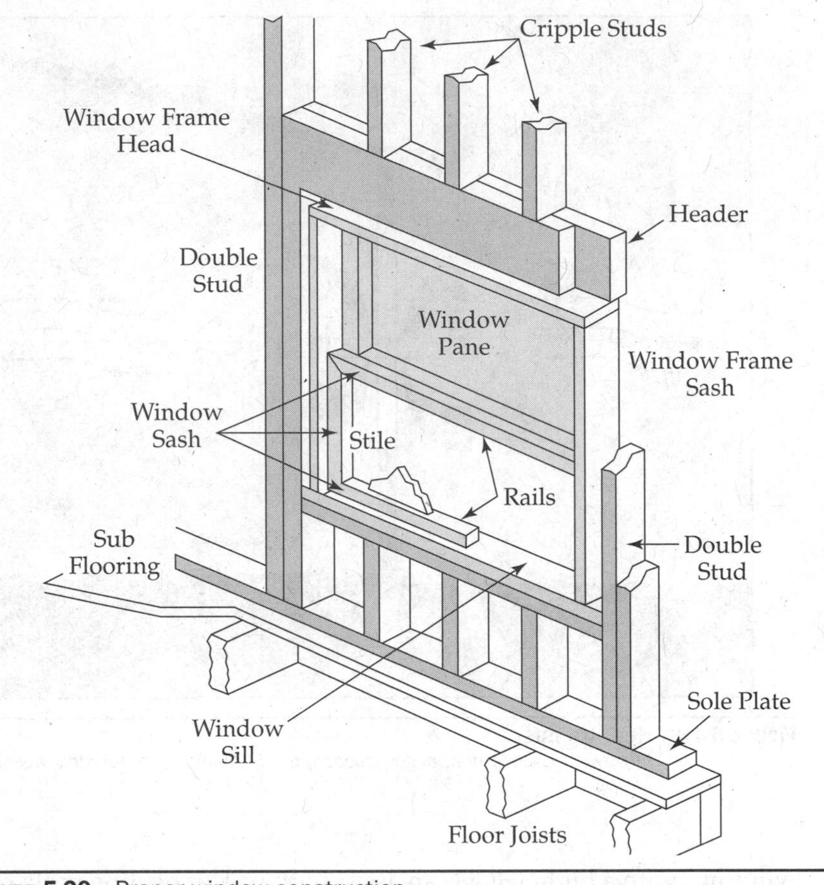

FIGURE 5-20 Proper window construction
http://www.cdc.gov/NCEH/publications/books/housing/Graphics/chapter_06/Figure6.07.jpg

ity to insulate. The additional cost per window as compared to using a nitrogen fill is small. Krypton is a gas that has better insulating properties than argon; however, it is considerably more expensive and not normally used.

The property that all of these gases share is low reactivity. What this means for windows is less reactivity, less heat transfer, and less energy lost or gained. In your energy-efficient home, you are attempting to control the amount of energy that enters and leaves your home. These gases help create a stable environment.

Shade and Awnings

For most homeowners, a passive solar project can entail a small home improvement, such as adding or removing sources of shade and sunlight (Figure 5-23).

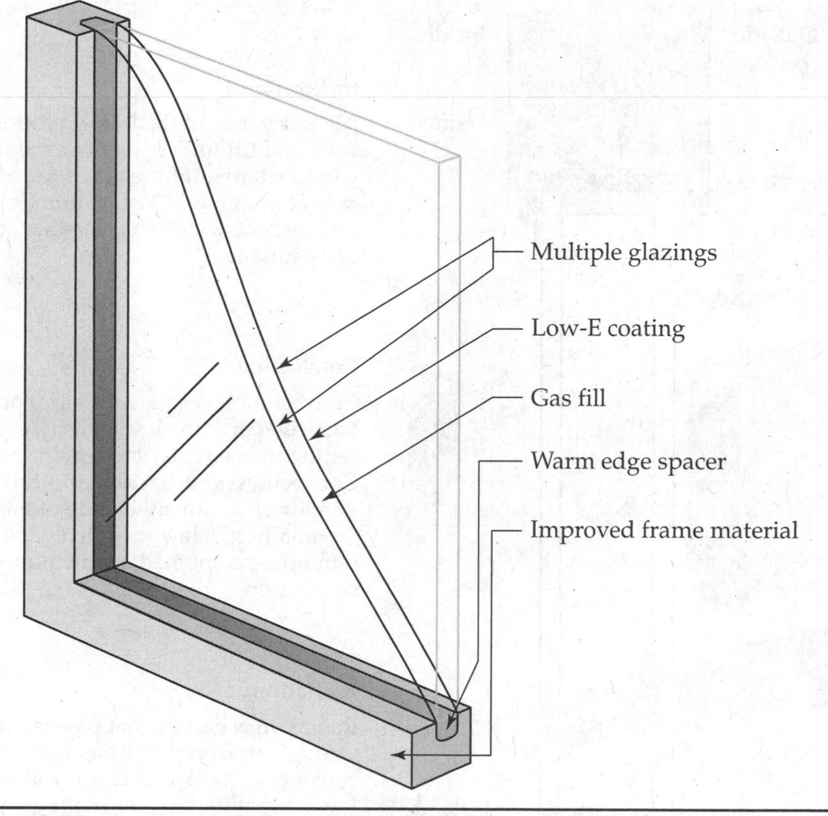

FIGURE 5-21 Gas-filled double-pane window
http://www.energystar.gov/ia/business/small_business/sb_guidebook/window.jpg

For small projects, one control mechanism is required: something to block the sun when it's unwanted or unneeded. External controls include shades and awnings (Figure 5-24), which are low cost, decorative, and easy to install.

A second external or internal option for shade is shutters (Figure 5-25).

Dual-pane, insulated windows are now available with blinds built in between the two panes of glass (Figure 5-26). Similar skylights are also available.

Landscaping

Another external way to use passive solar energy is with landscaping (Figure 5-27). You may not be able to reorient your home physically, but

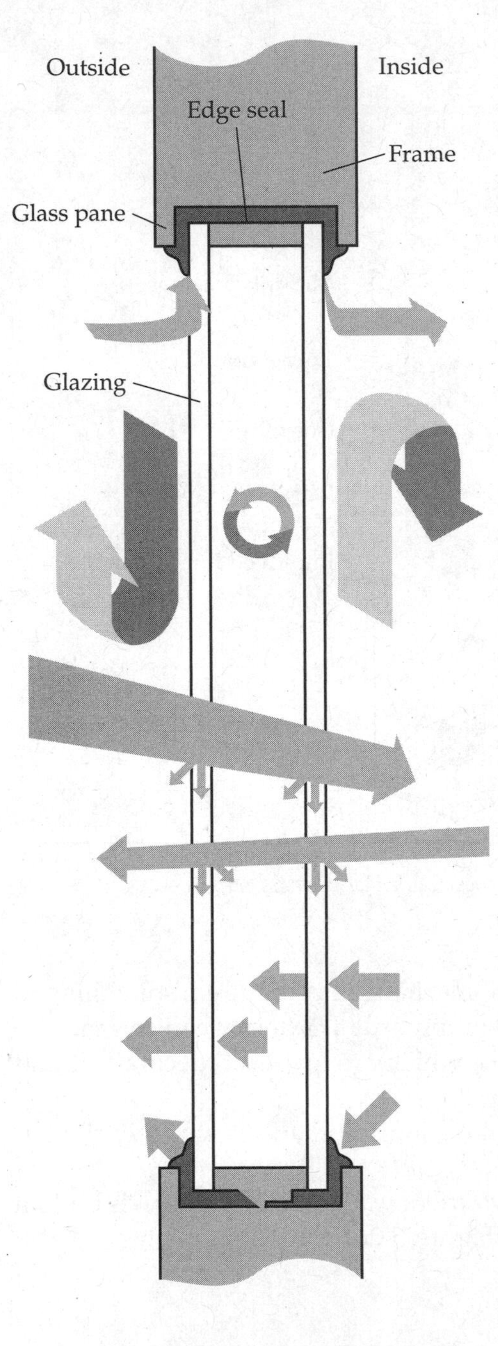

Infiltration

Air leaks around the frame, around the sash, and through gaps in movable window parts. Infiltration is foiled by careful design and installation (especially for operable windows), weather stripping, and caulking.

Convection

Convection takes place in gas. Pockets of high-temperature, low-density gas rise, setting up a circular movement pattern. Convection occurs within multiple-layer windows and on either side of the window. Optimally spacing gas-filled gaps minimizes combined conduction and convection.

Radiation

Radiation is energy that passes directly through air from a warmer surface to a cooler one. Radiation is controlled with low-emissivity films or coatings.

Conduction

Conduction occurs as adjacent molecules of gases or solids pass thermal energy between them. Conduction is minimized by adding layers to trap air spaces, and putting low-conductivity gases in those spaces. Frame conduction is reduced by using low-conductivity materials such as vinyl and fiberglass.

Courtesy: E SOURCE

FIGURE 5-22 Window performance
http://www.energystar.gov/ia/business/Web_art/EPA-BUM-SupLoads_7-2.gif

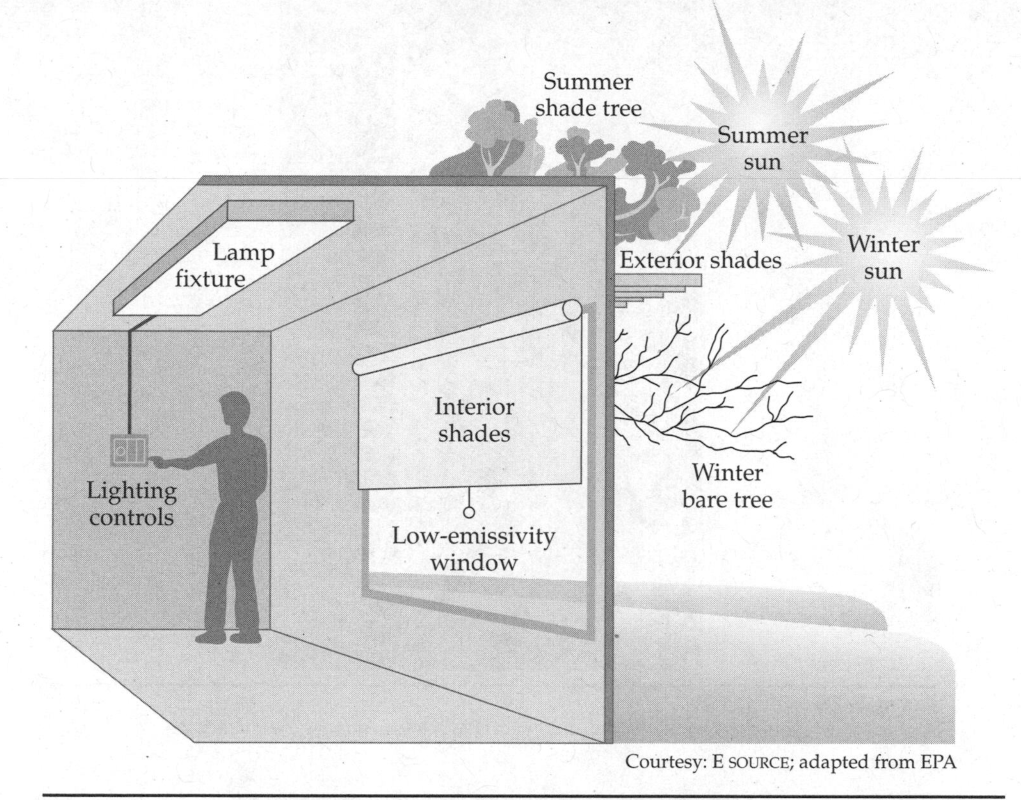

Courtesy: E SOURCE; adapted from EPA

FIGURE 5-23 Improving solar efficiency
http://www.energystar.gov/ia/business/Web_art/EPA-BUM-SupLoads_7-4.gif

FIGURE 5-24 Awnings
http://www.ci.la-porte.tx.us/images/3139_372.jpg

FIGURE 5-25 Shutters
https://www.eere-pmc.energy.gov/PMC_News/images/darmstadt.jpg

FIGURE 5-26 Window blinds between the glass panes
http://windows.lbl.gov/projects/DynamicShades/Image8.JPG

you can change the solar picture in your home by changing the landscaping. You can plant a tree to shade your home, and you can plant shrubs to reduce wind. Best of all, anything you grow will sequester carbon and produce water and oxygen.

Planting deciduous trees in a temperate climate is an excellent option for affecting the heating and cooling of your home. Deciduous trees will shade your home to keep it cool in the summer. In the winter, the same trees lose their foliage and allow sunlight to shine on your home (Figure 5-28).

Real estate agents know that a home with large trees sells for 15 to 20 percent more than a comparable home without trees. Who doesn't want to live in a beautiful community with tree-lined streets?

Solar and energy-efficient designs promise more comfort, better health, less cost, and no damage to the environment: the new American dream.

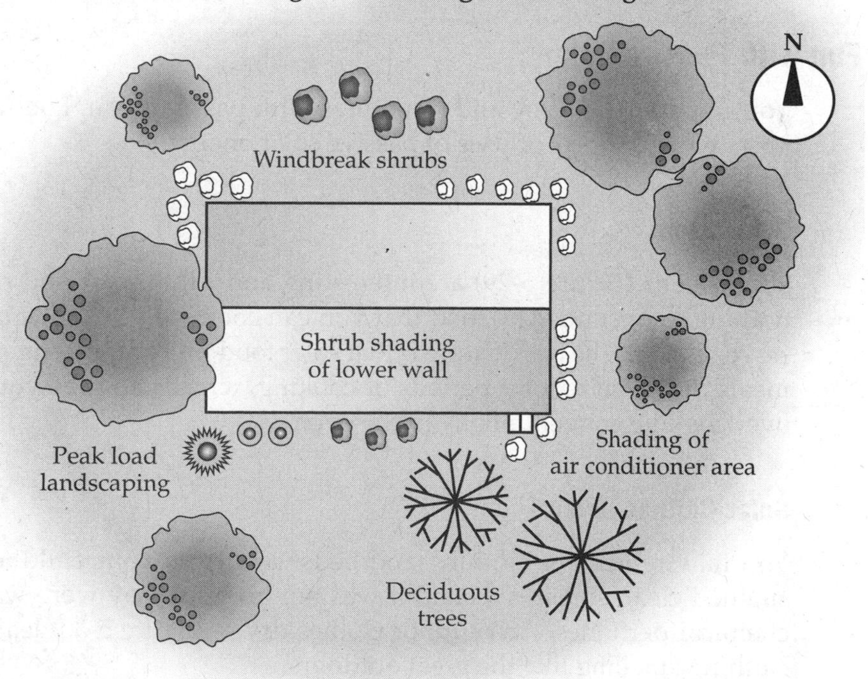

FIGURE 5-27 Landscaping for energy efficiency
http://www1.eere.energy.gov/buildings/residential/images/site_design_3.jpg

FIGURE 5-28 Deciduous trees for cooling and heating
http://www1.eere.energy.gov/consumer/tips/images/pg15_landscape.jpg

Fun with Passive Solar

You can do many easy and fun things with passive solar. Following are just a few examples of uses of passive solar energy.

Solar Oven

Solar ovens (Figure 5-29) are interesting and fun to use, and they cost nothing to operate. Any food that you can cook, you can cook in a solar stove or oven. The best feature is that solar food preparation complements meals that require long periods of cooking, which can save you money over cooking conventionally in the oven.

Solar Clothes Dryer

You may remember clothing and bedsheets from your childhood that smelled clean and fresh—and it was not because they were washed in chemical perfumes. An outdoor clothes dryer (Figure 5-30) leaves your clothing smelling like the great outdoors.

FIGURE 5-29 Parabolic solar oven
http://www.fnal.gov/pub/today/images09/ScienceAdventuresoven.jpg

FIGURE 5-30 Outdoor dryer

Solar Water Purification

Solar disinfection (SODIS) water purification (Figure 5-31) is a simple but effective process, because the sun's ultraviolet light destroys harmful bacteria in water. After filtering the water through ordinary sand, it is poured into cleaned plastic bottles and kept in direct sunlight for 4 to 6 hours, or longer if the water or sky is cloudy, to render the water clean and drinkable.

Boats often carry what is called a "solar still" for survival. A solar still functions on the premise of evaporation and condensation. A version of a solar still can be built on land, too (Figure 5-32): Dig a hole in the ground near a source of moisture, or add undrinkable water to the hole. Place a collection cup in the center of the still. Cover the hole with plastic, and place a small weight in the center of the plastic over the cup. Condensation will occur and drip down into the cup to produce clean, fresh drinkable water.

Passive Solar Cooling

Passive solar cooling may sound odd, but it is as simple as it is true: Heat rises, and cool air sinks. We can use this to design or modify buildings to use this moving air (Figure 5-33) to cool the building.

FIGURE 5-31 SODIS water disinfection in Indonesia
http://upload.wikimedia.org/wikipedia/commons/6/67/Indonesia-sodis-gross.jpg

FIGURE 5-32 Solar still
http://www.ars.usda.gov/aboutus/50th/historiesphotos/PacificWest/images/Original%20Files/JacksonSolarStill.jpg

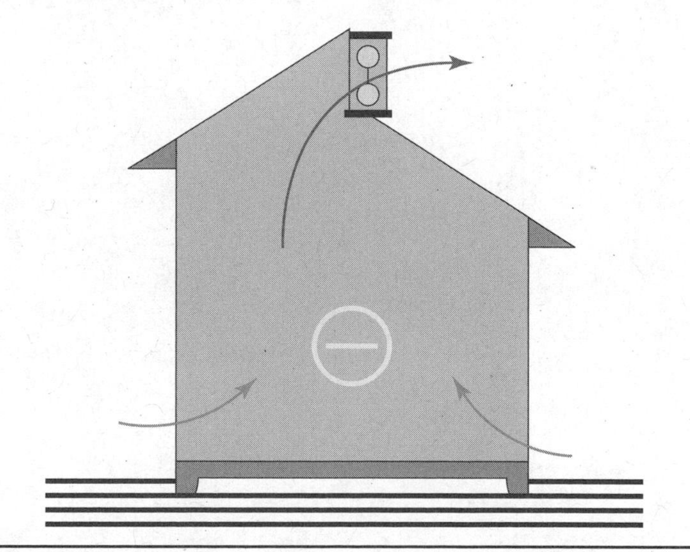

FIGURE 5-33 Passive cooling
http://dnr.louisiana.gov/SEC/EXECDIV/TECHASMT/ecep/drafting/c/dr-c24b.gif

Gardening

How about solar agriculture? Grow a garden and use the solar energy delivered free to your garden to grow food for you and your family.

CHAPTER 6

Creating a Personal Energy Plan

Solar is only part of a total home energy system. Installing a photovoltaic (PV) system and using other forms of solar energy should be coupled with a reduction in electricity and energy use. If you've read this far, you know that you can save more energy by choosing Energy Star–rated appliances and by making other modifications to your home and lifestyle.

A typical homeowner unknowingly wastes from 20 to 40 percent of the electricity consumed by failing to turn off lights and other appliances when they are not in use. Electricity is the easiest form of energy to waste, but it is also the easiest form of energy to save (Figure 6-1).

Using passive solar and reducing your energy usage will help reduce your utility bill and make your home more energy-efficient. Increasing your home's efficiency and improving energy use by choosing Energy Star–rated appliances will also contribute to your savings. Adding a PV system to your home can also help reduce your dependence on the utility company. But there is much more to the energy story. The best thing you can do to make your home energy use more efficient, in addition to the factors just mentioned, is to create a personal energy plan to manage your energy use.

Creating a total energy efficiency plan is the key to achieving real energy savings in your home. Taking a "total home" approach to saving energy ensures that the dollars you invest in energy are spent wisely.

Creating a Home Energy Plan

Creating a home energy plan is an important part of your energy-efficient home improvements. A viable plan will save you time, money, and energy.

FIGURE 6-1 Turn off the lights
http://www.cpsc.gov/cpscpub/prerel/prhtml10/10014.jpg

Begin by creating a strategy to deal with the largest energy consumers in your home. For most of us, this is the heating system (Figure 6-2). Then research and determine the most efficient solutions.

Start by creating a clear, well-defined energy goal. If your goal is energy improvement, home improvement, or financial improvement, then you've begun at the right place. A strategic plan is important whether you are developing a green renovation or simply replacing some energy-inefficient systems with energy-efficient systems in your home.

Here's how you create a strategic plan:

1. Ask an energy expert to evaluate your home's energy efficiency to identify the most significant energy drains in your home. This expert can help you define goals for energy efficiencies.
2. After you have defined your goals, write them down as succinctly as possible. If the goals or changes involve your family, include them in the plan.
3. Determine your budget, and be realistic. Keep in mind that most home projects often cost more than you first anticipated, so consider that as you plan. Do not change your budget to meet your plan; change your plan to meet your budget. Obtain cost estimates on the project to help determine whether your budget is reasonable.
4. Assemble information about any projects and activities that are required to meet your goals.
5. Determine what actions need to be accomplished and the order in which all activities must proceed. (Do not install the new carpet before you paint, for example.) A discussion with an expert may prove

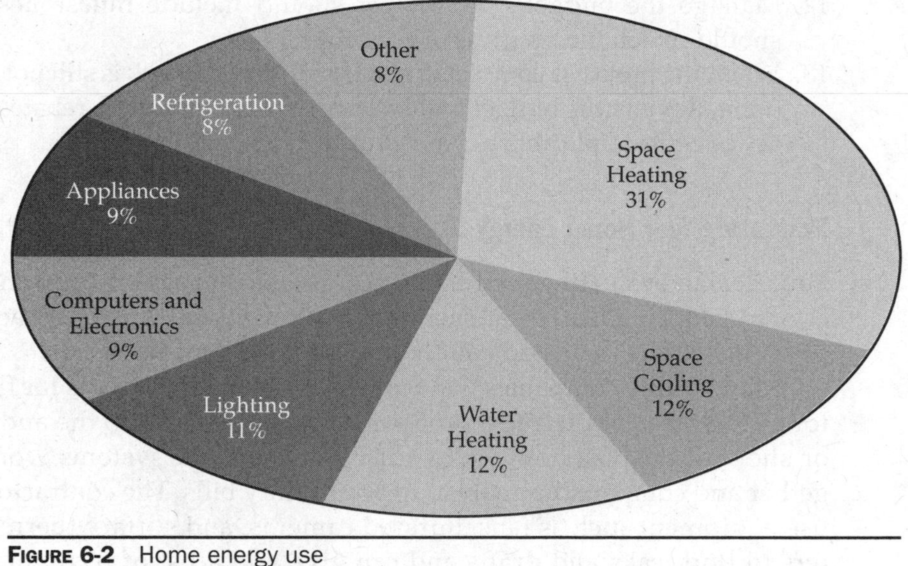

FIGURE 6-2 Home energy use
Source: Department of Energy 2007
http://www1.eere.energy.gov/consumer/tips/home_energy.html

useful. This is the time to begin putting the project plan, step by step, on paper.

6. Revise the plan as necessary. Synchronizing the budget and plan will help avoid fiscal problems as you move forward. Compare the budget and plan, and be sure that they are complementary.
7. Establish a timeline for each activity. If you're involving a contractor, consider his or her average completion time estimates, but know the worst case scenarios as well. Plan on the average timeline but be prepared to live with a project that takes more time than expected.
8. Establish milestones or target dates for work completed. Meeting milestones demonstrates progress. Milestones can help ensure that activities are completed prior to starting new ones. Some activities can or will occur concurrently. Milestones will not allow certain projects to begin until other milestones have been met.
9. Employ the labor required to complete your goals. Confirm who will be available to help complete the project.
10. Begin work on the project. Review the plan daily or as often as necessary. Revise work to meet the plan and budget, as needed. If the plan was properly designed, you should not need to revise it. Altering the plan once work has begun can lead to a significant increase in expenses.
11. Manage the project. Even if a contractor is employed, you are the manager of the project. Project and employee supervision will be a necessary daily task.

12. Manage the budget. The budget should include milestones and should match the progress of the project.
13. When the project is completed successfully, your work is still not quite done. Review the budget and be sure to collect refunds, rebates, and tax benefits applicable to your project.

Evaluating Your Home Energy Use

Your goal in evaluating your home energy use is to save money and energy by reducing inefficient energy use in your home. How you accomplish this goal depends on your home and your personal needs.

Many utility companies will perform home energy audits for free or for a small fee. If you hire a professional contractor to do the audit, he or she will analyze how well your home's energy systems work together and compare the analysis to your utility bills. The contractor can use equipment such as fans, infrared cameras, and surface thermometers to find leaks and drafts and can give you a list of recommendations for cost-effective energy improvements. A reputable contractor can also calculate the return on your investment in high-efficiency equipment compared with standard equipment. In some situations, a contractor performing the improvement can actually save you time and money.

In beginning to assess energy use, first gather all your utility bills. Write down the total of each bill for the year to determine the following:

- How much money is spent on each utility bill?
- Which expense is the largest?

Then, consider the following:

- Will you hire a contractor to do the work to fix the energy inefficiencies or improve your home, or will you do the work yourself? Some rebate and tax credit programs require that a certified professional do the work. If you want to do the work yourself and you want the rebates and credits, check this out before you start.
- How much will the work cost?
- How much of the work can you afford to do now?
- How much time will the work take?
- How much rebate or tax credits will be available after you do the work?
- How long will it take to recover the money spent to repair the problem?
- How long do you plan to own your current home?

In most energy audits, the heating and cooling expenses and energy use are often too high because of insufficient insulation in the attic and/or crawlspace. This particular problem is a good problem, because it's straightforward to fix, and product rebates, tax rebates, and tax credits can help you pay for the additional insulation and work.

How to Use the Plan

As you move through completing the steps necessary to reach your goal, you can make improvements to the home as they make sense. For example, you can install insulation in the walls at the same time that windows and door are to be installed. Having a clear plan of the total home system can help you consolidate the work done and will save time, money, and potential problems.

Having a plan before you begin can also help you make smart purchase decisions when buying items to use in your home. For example, if new windows are part of the home improvement plan, you can watch for sales on these products and purchase them before you begin the project.

After the job has been completed, you need to confirm that the insulation and other materials have actually accomplished your goals. Most tax credit and rebate programs require that you have your system retested after duct work is completed, insulation is added, or new windows and doors are installed. Your installer should be able to help provide the information and forms you need to apply for the tax credits and rebates.

What to Include in a Complete Energy Plan

Home electronics and home office equipment should also be included in a home energy assessment and plan. These items individually do not consume a significant amount of power, but home entertainment appliances, in particular, can be energy inefficient for the following reasons:

- These items are not usually purchased with energy savings in mind.
- Modern homes contain hundreds of such items.
- Most home entertainment appliances consume "vampire power" (standby power consumed when the device is not actually in use).

Following is a short list of small electrical products that you may use in your home:

- Battery chargers

- Cordless phones
- Cell phones
- DVD player
- Cable digital converter box
- Audio system: radio, stereo, iPod, CD player, receiver, speakers
- Home theater systems and sound equipment
- Power adapters
- Televisions
- Computers
- Copy and fax machines
- Digital duplicators
- Notebook computers/tablet PCs
- Monitors/displays
- Printer, scanners, and all-in-ones
- Router
- Scanner
- Digital camera
- Video camera

When you purchase these items, plan for shared use. For example, DVD player, home entertainment systems, audio systems, and home office products can be centralized so that all family members can access these appliances without duplication. You can load DVDs and CDs onto your computer so that all family members can easily access them any time they want. This saves space and energy. By consolidating equipment, the average homeowner can remove 10 or more redundant devices from the home.

A power strip saves money by allowing the user to turn off appliances that are not normally shut off.

When not in use, all these items should be turned off completely. Other energy-saving tips include the following:

- Charge your cell phone, laptop, iPod, and other portable items in the car. This electricity is almost free.
- When purchasing office equipment, purchase all-in-one devices (Figure 6-3) that combine fax, copier, printer, and scanner.
- Ask your cable or other service provider for energy-efficient set top boxes.
- Purchase an Energy Star–qualified cable and Internet-ready television monitor. Why have a TV and a large computer screen in the same room?

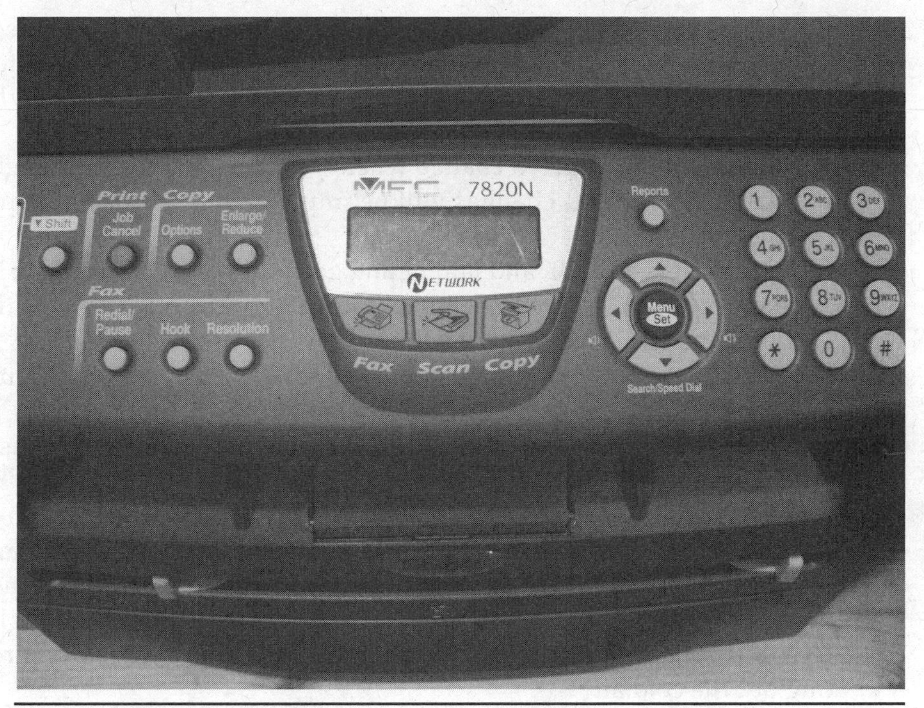

FIGURE 6-3 All-in-one device

Making the Most of Your Plan by Changing Old Habits

Energy saving in the home involves more than just replacing energy-consuming appliances and adding insulation; it's also a function of homeowner habits. Bad habits regarding energy use can cost you money and waste a lot of energy.

Habits, however, can be difficult to change. It can help to come up with a plan regarding how you can change, and then write it down, as you did with your home energy plan. You can refine or modify your plan as needed as you go. Your goal is to change old behaviors to more conscientious and conscious behaviors. Many factors in life can disrupt the best-made plans; what is important is how each person tries to adapt to the changes.

Once you've created and defined an ultimate goal, do the following:

- Develop milestones and "micro-plans" and allow for positive reinforcement when you've changed to new habits. Micro-plans can be added, modified, or eliminated as required without the entire plan changing.

- Perform good behaviors over and over again until they become new habits. It's more difficult to change a lifetime of habits without offering a new activity to replace that undesired habit. Replace the old habit with a new tradition.
- Make a commitment to change whenever you can. Try to change the plan or accept that changes will be implemented at the appropriate time.
- Ask partners, friends, and family, and get your family's buy-in on habit changes. They can help remind you to stay on track, especially if they're making changes as well.
- Write down motivating statements and read them every day. Maintain your goals by reading your written plan every day.
- Understand the causes of the original behaviors. Avoid situations that will cause the unwanted behavior.

When changing your habits, realize that bad habits are partly unconscious and mechanical. This is why behavior and thought patterns are not easy to change. Creating new patterns of successful behavior is a long-term but beneficial process. Now close the refrigerator door—you're letting out the cold air!

CHAPTER 7

The Fundamentals of a PV System

Photovoltaic (PV) solar panels (Figure 7-1) convert light energy to electricity.

PV solar is most effective when it's combined with other factors in an energy-efficient structure. Adding passive solar, compact florescent light bulbs, appropriate amounts of insulation, and energy-efficient appliances complement the benefits of any PV system.

PV solar can be and is used almost everywhere. These systems are stable, reliable, pollution-free, affordable, and profitable if you can sell excess power to a utility company. As the production of PV panels increases and PV solar becomes more popular, the cost of PV solar is becoming more and more affordable.

Read *Do-It-Yourself Home Energy Audits: 140 Simple Solutions to Lower Energy Costs, Increase Your Home's Efficiency, and Save the Environment* by David S. Findley (McGraw-Hill, 2010), before you add a PV solar system to your home.

Solar is an intermittent power source that functions only when the sun is shining, but even with this limitation, even in cloudy parts of the world, such as Seattle, PV power can be used successfully. Germany is comparable to the Pacific Northwest in number of days per year of sunlight, with approximately 130 days, and Germany has a very successful solar program. In addition, the northeastern United States receives approximately 230 days a year of sunlight, and the Southwest gets 300-plus days per year. In fact, most areas of the United States receive enough sunlight to make use of solar energy (Figure 7-2).

FIGURE 7-1 PV solar panels in Aspen, Colorado
https://www.eere-pmc.energy.gov/PMC_News/images/Aspen-solar.jpg

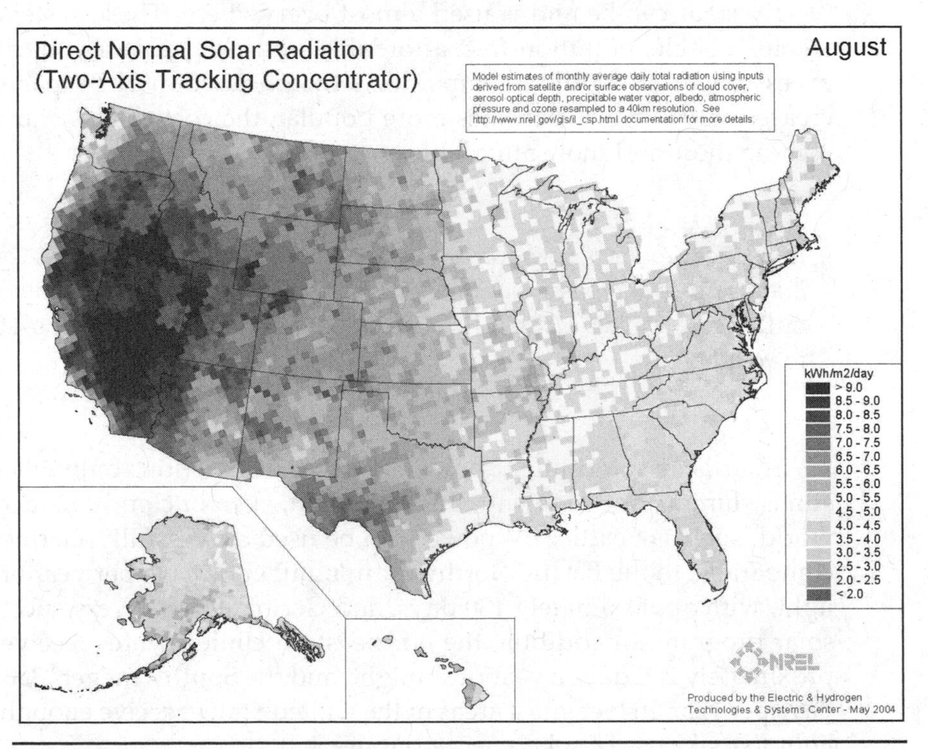

FIGURE 7-2 Map of solar radiation
http://www.nrel.gov/gis/images/map_csp_us_august_may2004.jpg

To make PV systems even more affordable, several states offer financial incentives through solar rebates and other programs. Some utilities offer net metering programs that allow your PV system to return the excess electricity produced to the utility grid—you can even sell the excess electricity your system produces to the utility grid for a profit.

I recommend that you consult an experienced PV system designer and installer. A PV system is a substantial investment and can be a complicated process. In addition, many rebates and tax deduction programs require that you use a licensed local installer.

PV panels are mounted and oriented so that they can catch the most sunlight. Because more electricity is produced when the sunlight is more intense and/or strikes the PV modules directly at a perpendicular angle (Figure 7-3), all solar panels are angled for optimum solar collection.

As you can see in Figure 7-4, the angle at which your PV system is installed depends on your latitude. Panel angles are adjusted to make sure the sun shines as perpendicular to the panels as possible.

The season is another important factor in how the panels are angled. The sun is in different portions or elevations in the sky depending on the time of year (Figure 7-5).

Figure 7-3 Solar panel angle
http://www.ia.nrcs.usda.gov/news/images/Pics/solarpanels4621.gif

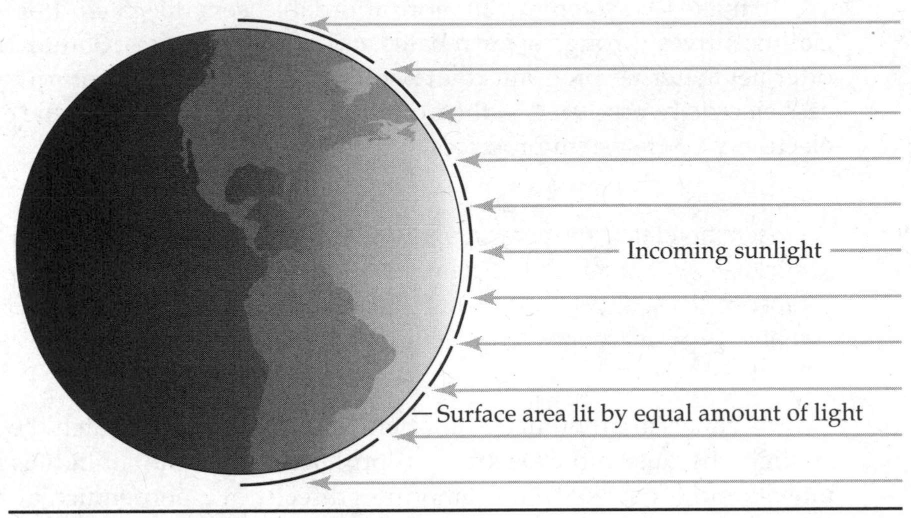

Figure 7-4 Sunlight angle of attack
http://earthobservatory.nasa.gov/Features/EnergyBalance/images/sunlight_angle.png

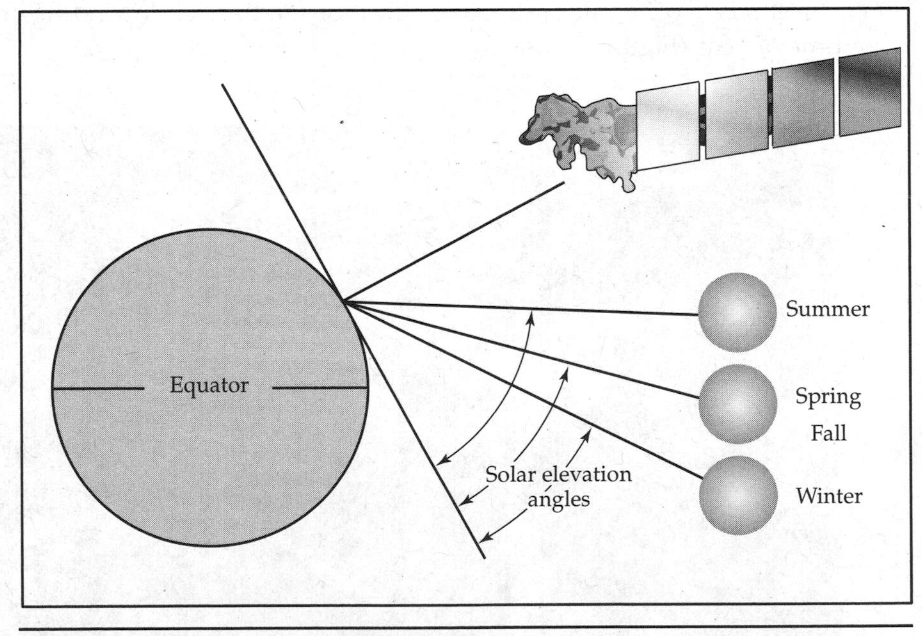

Figure 7-5 Sun seasonal elevation changes
http://landsathandbook.gsfc.nasa.gov/handbook/handbook_htmls/chapter6/images/sun_elevation_changes.jpg

PV System Components

The basic element of a PV system is the solar cell (Figure 7-6), which is often made of *silicon*, a semiconductor material that is commonly used in solar cells and computer chips. Other cell materials are *polycrystalline* thin film and single *crystalline* thin film (all discussed later in this chapter). Multiple solar cells are connected together to form a PV *module*. Multiple modules are connected to form solar *panels*, and multiple panels form a solar *array* (Figure 7-7). The electrons freed by the interaction of sunlight with semiconductor materials in PV cells are captured to create an electric current. Each solar cell produces 0.5 volt, each solar panel produces 180 to 300 volts, and a solar array production is limited only by the number of panels that are connected.

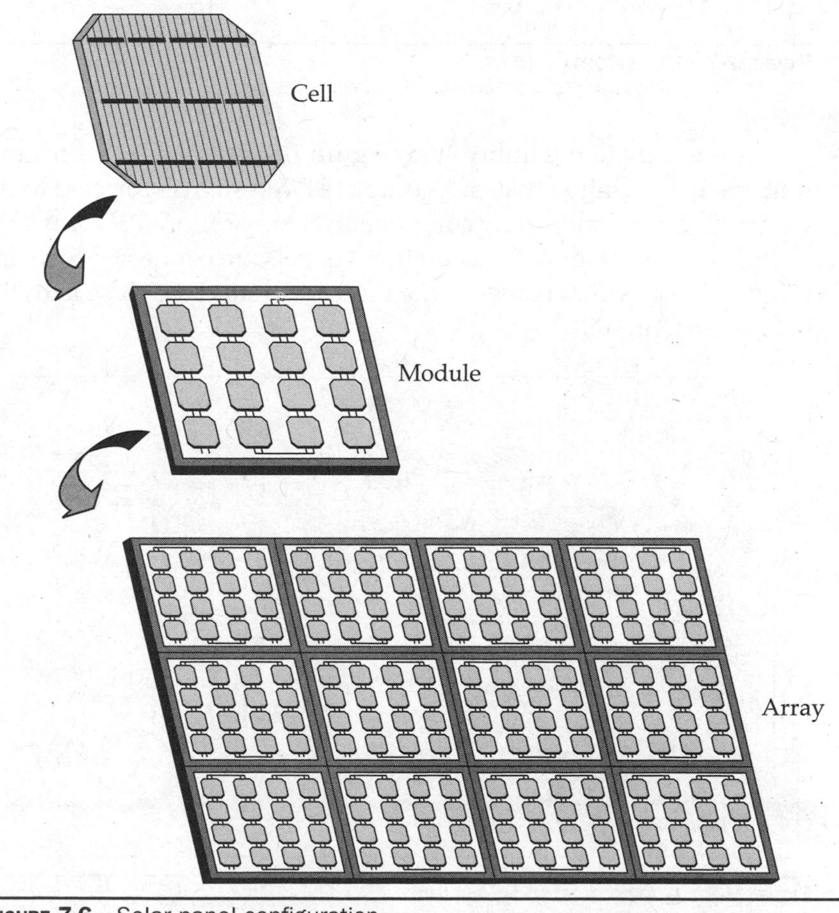

FIGURE 7-6 Solar panel configuration
http://www.energyquest.ca.gov/story/images/chap15_array_nasa.jpg

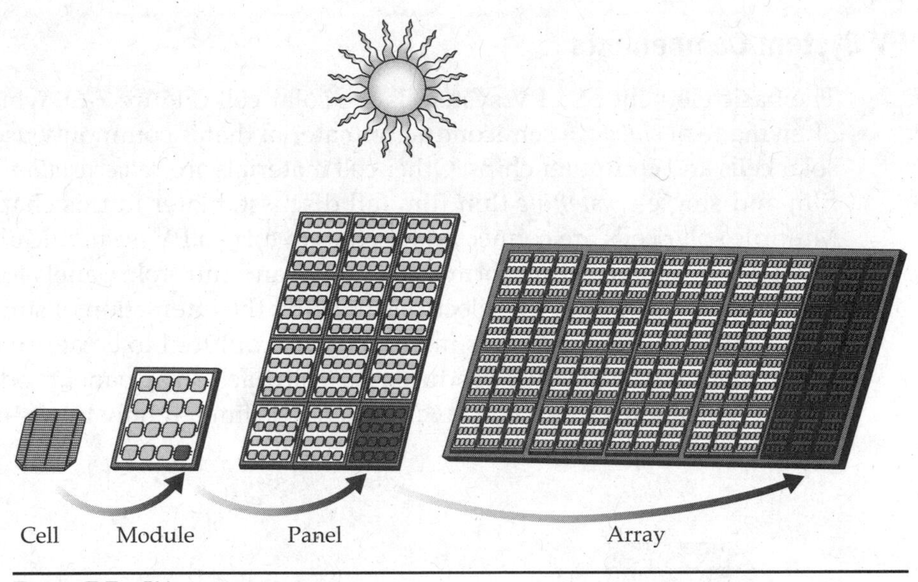

FIGURE 7-7 PV system
http://www.zeh.ca/Portals/0/Pictures/Cell-Array.JPG

Connecting to the utility grid (Figure 7-8) is a possibility for most PV systems in the United States. A complete PV system connected to the utility grid requires only a few components. One or more PV arrays are connected in a rack (Figure 7-9), and these panels are connected to an inverter (Figure 7-10), which converts the direct current, DC, electricity to alternating current, AC.

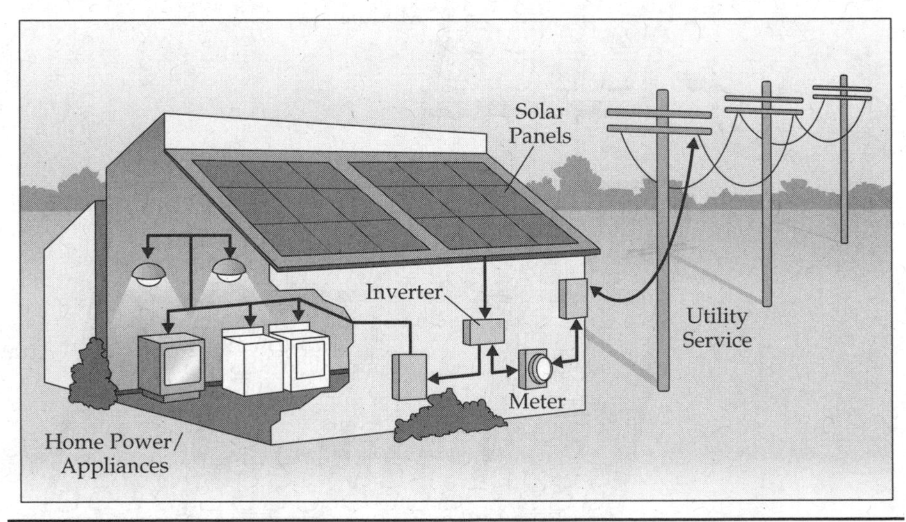

FIGURE 7-8 Grid-connected PV system
http://www.energysavers.gov/images/residential_grid_pv.gif

Figure 7-9 Ground-mounted PV
http://www.truewest.ca/images/Inverter_Armstrong.jpg

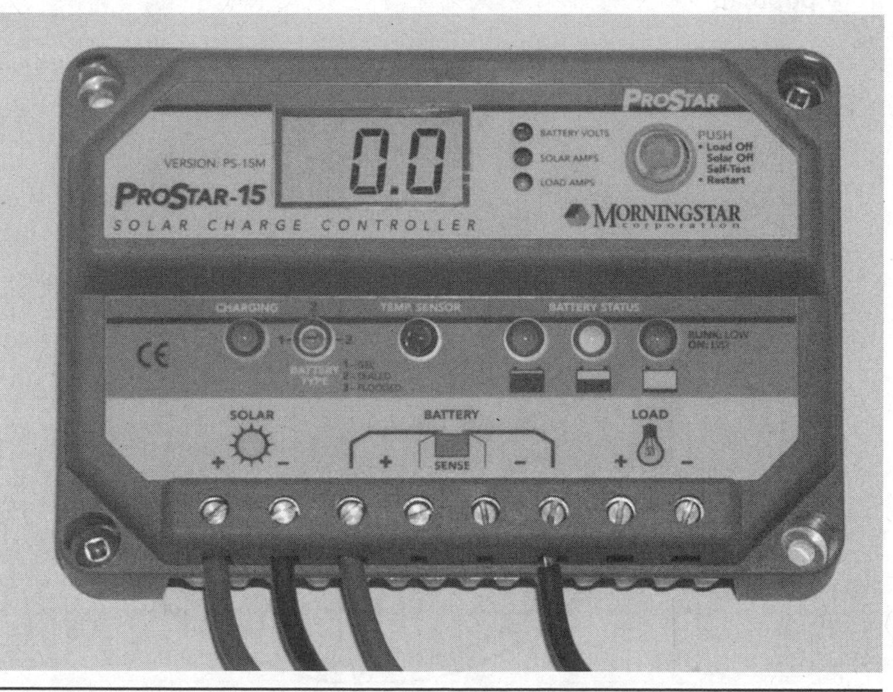

Figure 7-10 PV controller
http://www.cassidyautomation.ca/_2173624_edited-1.jpg

The inverter is connected to a controller if batteries are used. The controller directs the flow of electricity. The controller is then connected into a power meter that measures electrical flow in both directions (Figure 7-11). This is called *net metering*, and this is what allows you to send or sell surplus electricity to the grid. Batteries are optional to provide energy storage or backup power in the event of a power interruption or an outage from the grid.

Even if you live in a remote area, stand-alone PV systems with batteries can be purchased to store energy for later use. This type of PV system is common and easy to implement and use.

The type of system you choose is usually dictated primarily by cost. Backup systems, including batteries, generators, or hydrogen systems, can be used at an additional cost that most homeowners are unwilling to accept. Hybrid systems with fuel cells or generators are also possible, but these systems are significantly more expensive than a simple PV project.

This book focuses primarily on simple PV systems. Backup and hybrid systems are discussed, but because they are cost prohibitive and uncommon (Figure 7-12), they are not covered extensively. Finding an installer that can provide a complete hybrid system can also be difficult, though that could change in the future as these systems become more popular.

FIGURE 7-11 Electrical meter
http://www.lakelandpower.on.ca/Portals/1/Pictures/meter.bmp

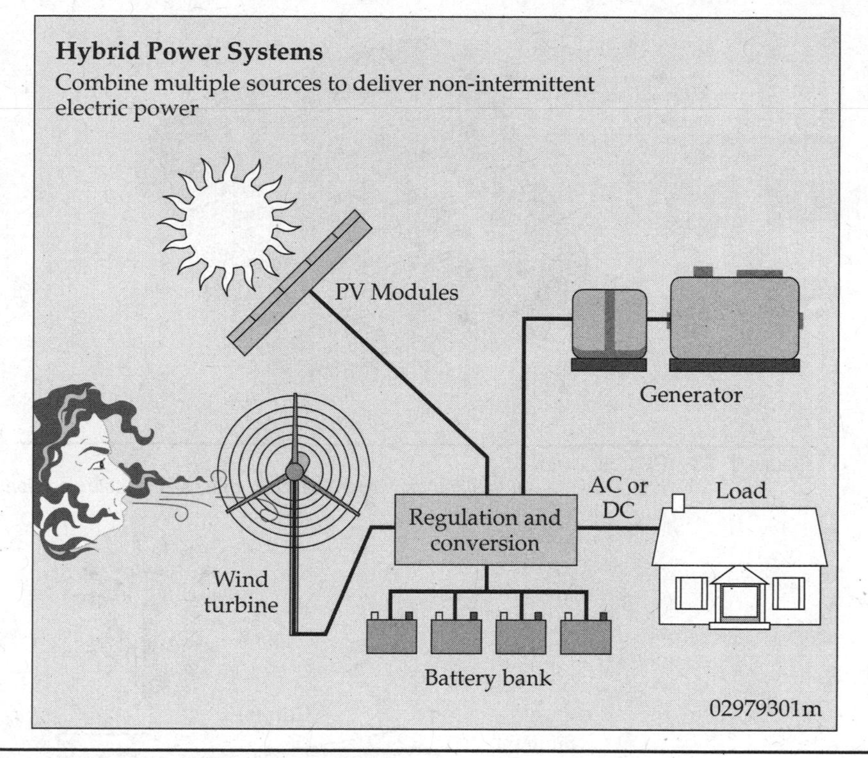

FIGURE 7-12 Hybrid PV system
http://www1.eere.energy.gov/buildings/residential/images/wind-powered_electric_systems_3.gif

How PV Cells Work

A typical solar PV cell consists of a glass or plastic cover or other encapsulant, an anti-reflective layer, a front contact to allow electrons to enter a circuit, a back contact to allow them to complete the circuit, and the semiconductor layers where the electrons begin and complete their journey.

You can think of PV cells (Figure 7-13) as individual power plants that convert solar light into electricity. When sunlight strikes the cell material, it releases electrons that flow along an electrical wire (Figure 7-14).

The light spectrum (Figure 7-15) is also important to a PV cell. The part of the spectrum used by the silicon in a PV module is from 0.3 to 0.6 micrometers, approximately the same wavelengths that the human eye can see. These wavelengths encompass the highest energy region of the solar spectrum. Traditional cells use only this short range of light energy.

Figure 7-13 PV cell
http://www.gosolarcalifornia.ca.gov/solar101/images/nrel_solar_cell.jpg

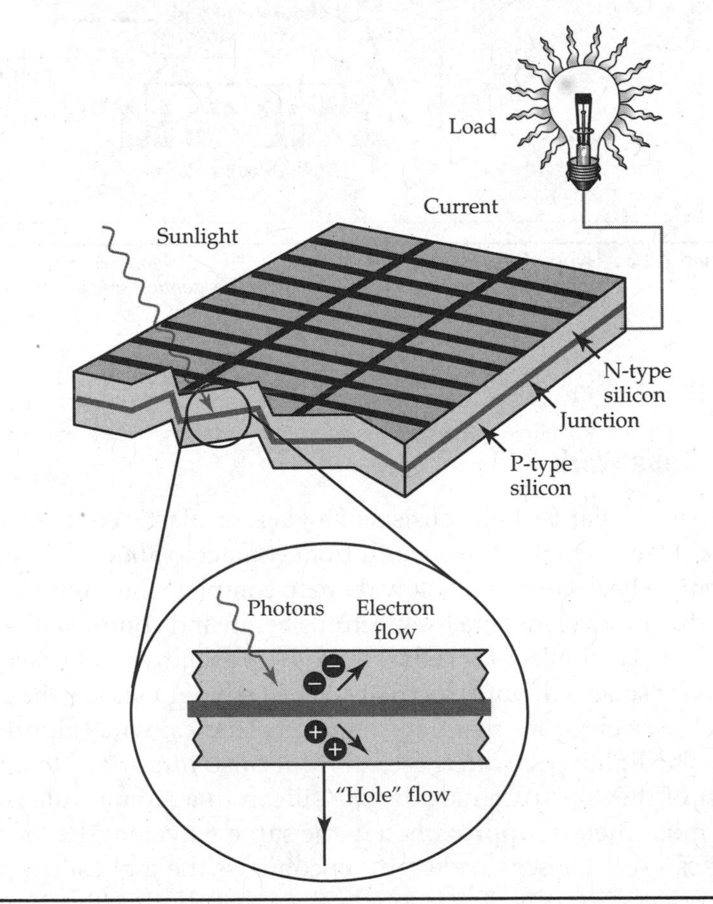

Figure 7-14 PV cell electrical production
http://www.energyeducation.tx.gov/renewables/section_3/topics/photovoltaic_cells/img/fig13photo-cell.png

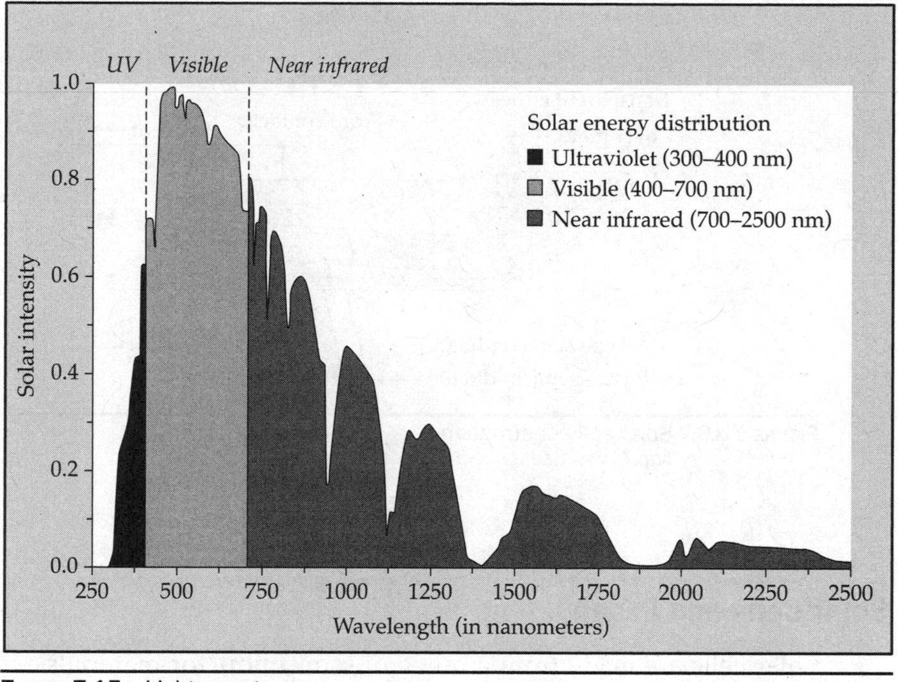

FIGURE 7-15 Light spectrum
http://www.lbl.gov/Science-Articles/Archive/sb/Aug-2004/reflectance_wavelength.jpg

The *absorption coefficient* of a material indicates how far light having a specific wavelength can penetrate the material before being absorbed. A small absorption coefficient means that light is not readily absorbed by the material. The absorption coefficient of a solar cell depends on two factors: the material making up the cell, and the wavelength or energy of the light being absorbed.

Another term important to the PV cell is the *bandgap*. The bandgap of a semiconductor material is the minimum amount of energy that is needed to move an electron from its static state. The lowest energy level of a semiconductor material is called the *valence band*. The higher energy level where an electron is free to roam is called the *conduction band*. The *bandgap* is the energy difference between the conduction band and valence band.

The semiconductor layer of the solar cell is where current is created (Figure 7-16). Creating this perfect layer is expensive, because any impurities will affect performance.

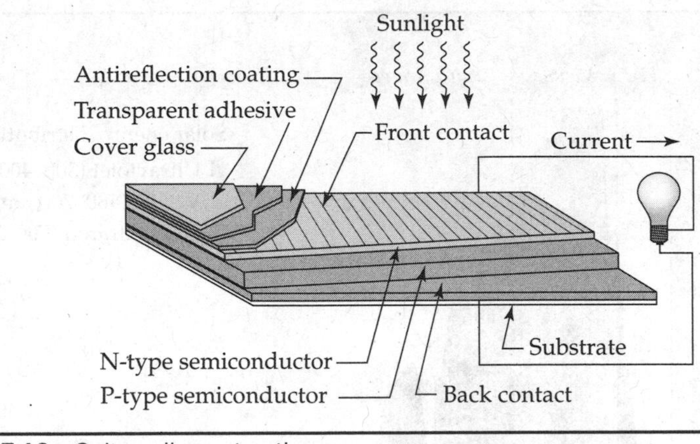

FIGURE 7-16 Solar cell construction
http://www1.eere.energy.gov/solar/solar_cell_materials.html

Solar Cells and Panels

Solar cells are made from a range of semiconductor materials—usually one of three types of base materials:

- Silicon (Si), which includes single-crystalline, multicrystalline, and amorphous forms
- Polycrystalline thin films, which include copper indium diselenide (CIS), cadmium telluride (CdTe), and thin-film silicon
- Single crystalline thin films, which include gallium arsenide (GaAs)

The type of material used and how it is formed is of great importance. The *crystallinity* of materials indicates how perfectly ordered the atoms are in the crystal structure. Silicon and other semiconductor materials come in these main forms:

- **Single or monocrystalline:** Crystals are repeated in a regular pattern from layer to layer.
- **Polycrystalline or multicrystalline:** Small crystals are arranged randomly, similar to shattered glass.
- **Amorphous silicon or thin film:** Materials in these panels have no crystalline structure.

For most practical purposes, all commercially available solar panel types function in similar ways. Your choice of panels depends on how much power you require, how much room you have you have for panels, and where they will be mounted.

Single- or Monocrystalline-Silicon Panels

Monocrystalline silicon panels use crystalline silicon, a basic semiconductor material. A "mono panel" has individual cells of silicon. The panels are produced in large sheets that can be cut to the size of a panel and integrated into the panel as a single large cell. Conducting metal strips are laid over the entire cell to capture electrons and form an electrical current.

Monocrystalline is the oldest technology and the most expensive to produce, but it offers the highest efficiency. Single silicon crystals are sliced into thin layers to make individual wafers. A solar panel is made up of a matrix of these wafers laid flat. They are long-lasting and degrade slowly.

Polycrystalline, or Multicrystalline, Silicon Panels

Polycrystalline silicon panels are the most common type of solar panels on the market today. Polycrystalline, or multicrystalline, PVs use a series of cells instead of one large cell (Figure 7-17). These panels consists of multiple small silicon crystals that look a lot like shattered glass. They are slightly less efficient than the monocrystalline silicon panels, but they are less expensive to produce.

FIGURE 7-17 Polycrystalline solar panels
http://www.bnl.gov/energy/images/Solar-300px.jpg

Although they are less efficient than single crystal, once the polycrystalline cells are set into a frame with 35 or so other cells, the actual difference in watts produced per square foot is not significant. These panels are also sliced from long cylinders of silicon, but the silicon used is multicrystalline, which is easier to make. They are similar to monocrystalline panels in performance and degradation.

Several production techniques can be used to create polycrystalline panels.

Cast Polysilicon

In this process, molten silicon is first cast in a large block to form crystalline silicon. Then the block is shaved across its width to create thin wafers to be used in PV cells. These cells are then assembled in a panel. Conducting metal strips are laid over the cells, connecting them to each other and forming a continuous electrical current throughout the panel.

String Ribbon Silicon

String ribbon PVs use a variation on the polycrystalline production process. Molten silicon is drawn into thin strips of crystalline silicon using metal strings. These strips of PV material are then assembled in a panel. Metal conductor strips are attached to each strip to create a path for the electrical current. This technology reduces costs and it eliminates the process of producing wafers.

Amorphous Silicon or Thin-Film Panels

Thin-film panels are quite different from crystalline panels (Figure 7-18). Materials in these panels have no crystalline structure. Instead, silicon is sprayed or applied onto a base as a thin film. The silicon materials are then connected to the same metal conductor strips used in other technologies. Amorphous panels do not require the same level of protection needed for more fragile crystalline cells during production or use.

Thin-film is a relatively new product, so its 20-year performance can only be estimated. One company guarantees less than 20 percent degradation over 20 years, which compares with 10 percent for the other types of panels mentioned.

The primary advantages of thin-film panels are the low manufacturing costs and versatility. The production process is more energy-efficient than that of the other cell types, so the cells are typically cheaper for the same rated power. Thin-film panels are less efficient, but amorphous silicon does not depend on the long, expensive process of creating silicon

FIGURE 7-18 Amorphous silicon or thin-film panels
http://www.nrel.gov/pv/thin_film/docs/shell_solar_cis_roof.jpg

crystals, so these panels can be produced more quickly and efficiently. Additional components are not required, so costs are reduced further.

Thin-film panels have several significant disadvantages. Production cost is low but so is efficiency. Thin-film technologies also depend on silicon, which has high levels of impurities. This reduces efficiency rapidly over the life of the product.

The big news in thin film is from some of the largest manufacturers. In 2010, two of the leading manufacturers increased production in exponential proportions. This may allow the sale of thin-film at half the cost, in energy production, of coal. This may mark the beginning of a thin-film solar revolution.

Each type of solar panel degrades at a different rate, so although some panels provide a great cost-performance ratio up front, you should compare the performance guarantee 20 years into the future. Thin-film solar panels generally degrade approximately 1 percent each year, whereas crystalline panels degrade at approximately 0.5 percent annually.

Group III and V Technologies

The future of PV solar is in Group III and V technologies. These technologies use a variety of materials with very high conversion efficiencies, claiming efficiencies of 25 to 80 percent.

These materials are categorized as Group III and Group V elements in the periodic table, hence their name. Gallium arsenide is combined with other materials to create semiconductors that can react to different types of solar energy. These technologies are very efficient; however, their use is limited due to the material costs.

Building-Integrated Photovoltaic

Building-integrated PV (BIPV), another category of promising new solar products, is designed to serve a dual purpose: it's a standard construction material that can also produce electricity. For example, it can be used as an amorphous silicon roofing material (Figure 7-19). BIPV technologies have very low efficiency levels due to their use of amorphous silicon. The advantage is that this material can be used anywhere, so that large areas can be devoted to producing electricity.

FIGURE 7-19 Building-integrated PV
http://www.nrel.gov/pv/thin_film/docs/shell_cis_bipv_dsc01285.jpg

Concentrated Solar Power

Concentrated solar power (CSP) systems are designed to increase efficiency of solar PVs by using "focused energy." CSP PV systems use a standard PV panel with concentrating lenses that gather sunlight. They require less silicon for the same amount of output and reduce the amount of space required for a PV installation.

The disadvantage of CSP is its reliance on direct light to produce electricity. PV panels can use both direct and diffuse light, but CSP PV systems require direct light. They do function well with tracking solar (discussed later in the chapter); however, many geographical regions do not receive enough direct light. The second disadvantage is that the lenses require maintenance to sustain efficiency.

High-Efficiency Multijunction Devices

High-efficiency multijunction devices (HEMDs) are an evolving technology that uses multiple layers of PV cells (Figure 7-20). Each layer acts as a link where solar energy is absorbed. Each layer in an HEMD is made from a different material with its own receptivity to each type of solar energy. HEMDs takes advantage of the entire light spectrum.

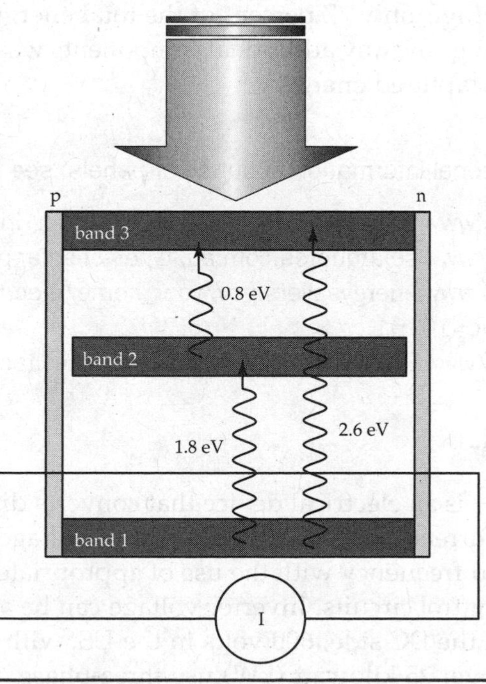

FIGURE 7-20 High-efficiency multijunction devices
http://emat-solar.lbl.gov/images/multiband_v02.jpg

In an HEMD, the top PV layer reacts to light that travels in short wavelengths and has the highest amount of energy. Each layer absorbs solar energy as the waves pass through the sheet.

Solar Panel Efficiency

A solar panel's energy conversion efficiency is the percentage of power converted (from absorbed light to electrical energy) and collected when a solar cell is connected to an electrical circuit.

The following numbers are the advertised percentages of efficiency for each of the different types of solar panels:

- Monocrystalline: 19 percent
- Polycrystalline: 15 percent
- Amorphous (thin-film): 10 percent

Unfortunately, the best solar panels, under ideal conditions, are about 19 percent efficient. This means that 81 percent of the energy that reaches your solar panel is not used. Of the 19 percent energy captured, under ideal conditions, the inverter then wastes 5 to 10 percent of that energy (so, on average, only 77 percent of the total energy is used). The electric meter, wiring, and any additional components waste more of the original 19 percent captured energy.

For additional information about solar panels, see these websites:

- http://www.masstech.org/cleanenergy/solar_info/types.htm
- http://www.solarquotes.com.au/types-of-solar-panel.html
- http://www.energysavers.gov/your_home/electricity/index.cfm/mytopic=10791
- http://www1.eere.energy.gov/solar/photovoltaics.html

The Inverter

An inverter is an electrical device that converts direct current, DC, to alternating current, AC (Figure 7-21). The AC voltage can be at any required voltage and frequency with the use of appropriate transformers, switching, and control circuits. Inverter voltage can be as high as 1000 volts in Europe on the DC side, 600 volts in the US, with no minimum voltage. Inverters over 25 kilowatt (kW) use three-phase, while most residential systems of 10 kW or less use single-phase.

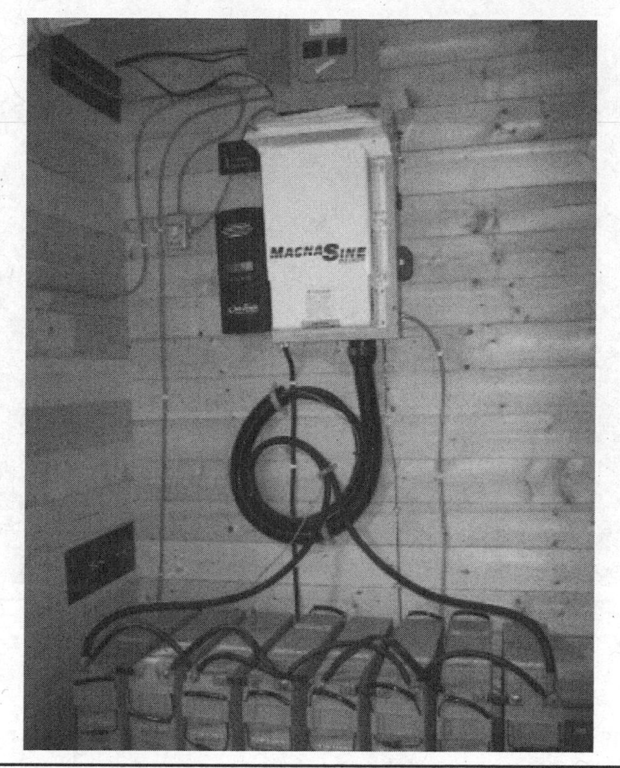

FIGURE 7-21 The electrical inverter
http://www.pinecottages.ca/images/Solar%20Power%20Pics/Battery-Inverter.jpg

Inverter efficiencies are currently between 85 and 95 percent, although recent developments have some large manufacturers claiming inverter efficiencies for solar power of 98 percent. Complementing your products with compatible components will increase efficiency of the overall PV system. Ask your solar supplier or installer about individual component and overall efficiencies.

Electrical Meters

When installing a PV system, if net metering is available, you will be required to install a new electrical meter (Figure 7-22). This net meter allows for a measurement of net energy consumption—both entering and leaving the system.

Solar Tracking System

Under ideal conditions, direct perpendicular sunlight is the most efficient for producing electricity. A roof-mounted PV system receives optimal con-

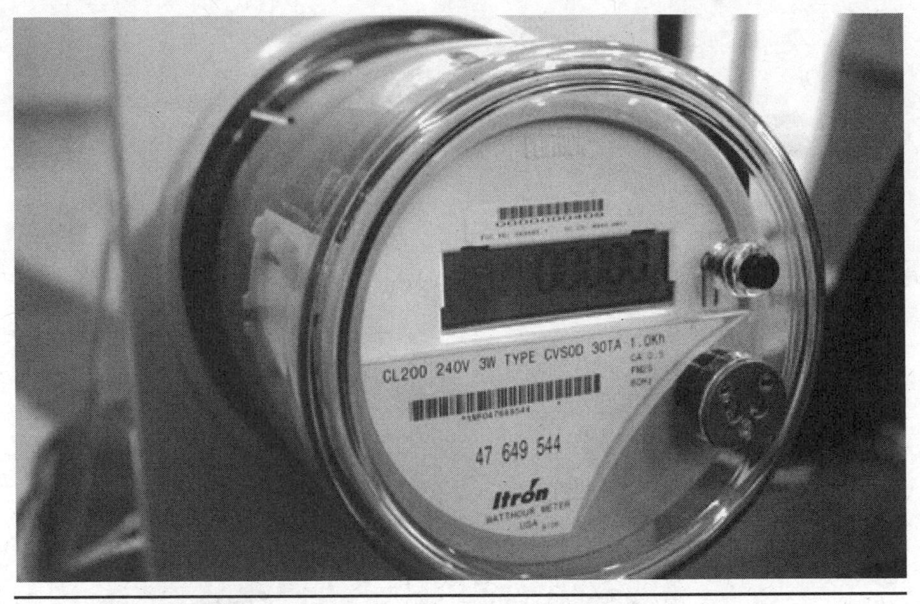

FIGURE 7-22 Electrical meter
http://www.cbc.ca/gfx/images/news/photos/2009/03/11/metre-cp-w-2318553.jpg

ditions for only a brief period during the day. A solar tracking system (Figure 7-23) can track the sun from sunrise to sunset and can create optimal power. A tracking PV system can also accommodate for seasonal changes of the sun.

FIGURE 7-23 Solar tracking system
https://newsline.llnl.gov/retooling/apr/images/04.25.08/PV_1.jpg

Some solar companies claim that as much as 50 percent more energy can be produced with a good tracking system over a fixed system.

The best option is a pole-mounted PV panel with a tracking system. A solar tracking system can be computer controlled to track the sun across the sky hourly and seasonally.

PV Energy Production and Savings

You can determine how well PV solar will work for you with a few calculations. Start by taking a look at your electric bill. I'll use my own bill as an example here to show you how this is done:

According to my monthly bill, I paid a total of 29 cents per kilowatt hour for electricity. My home uses about 500 kWh (or 500,000 watt hours) per month, or 6000 kWh per year. I can use this information to determine the most appropriate sized PV system to install.

500000 watt hours / 30 days = 16,667 watt hours used per day. I know that a 10 kW PV system will ideally produce a day's worth of power in 1.7 hours. I also know that a properly designed PV system will produce around 77 percent of its maximum output (see the section "Solar Panel Efficiency" earlier in the chapter). So a 10 kW system would require 2 hours of perfect sunshine every day to meet all of my requirements.

You can use average calculations based on your electric bill. A PV system installer should be able to offer a realistic average for any size system to meet your needs. I recommend buying the largest system possible according to available rebates and tax incentives. Most people use the majority of electricity that they produce, but if your PV system is too large for your home, you can sell excess energy to your utility company.

How Much Does a Typical PV System Cost?

A 10 kW system installed in a home can cost $30,000 to $40,000 or $6 to $8 per watt produced. These prices are rough estimates; your costs depend on your system's configuration, your equipment options, and other factors.

The following table shows sample costs for an average 5200 watt/5.2 kW PV system in the state of New York:

Installed cost (assumes $7.40 per watt)	$38,480
Rebate (at $3.50 per watt)	–18,200
Cost after rebate	$20,280
Less 30 percent federal tax credit	(6084)

New York state tax credit (lower of 25 percent or $5000)	(5000)
Total federal and state tax credits	–11,084
Final customer investment (based on output 6620kWh annually at 0.196 cents / kWh rate)	$9196

Source: Long Island Power Authority 2010
http://www.lipower.org/efficiency/solar-estimator.html

If you want to figure out how much energy savings you can count on from PV solar, you can do the following:

1. Based on your geographic location, select the energy production factor from the map shown in Figure 7-2 for the "kWh/kW per year" input for the equations.
2. Energy from the PV system = (kW of PV) × (kWh/kW per year) = kWh per year. Divide this number by 12 if you want to determine your monthly energy reduction.
3. Energy bills savings = (kWh per year) × (Residential Rate)/100 = $ per year saved. (Residential Rate in this equation should be in dollars per kWh; for example, a rate of 10 cents per kWh is input as $0.10/kWh.) For example, a 2-kW PV system in Denver, at a residential energy rate of $0.07/kWh, would save about $266 per year: 1,900 kWh/kW per year × $0.07/kWh × 2 kW = $266/year.

Rebate and Deduction Caveats

A PV system is a specialty home improvement, and qualifications for any rebates or tax deductions depend on your following specific installation instructions and filling out the appropriate forms. Consult with a local PV solar installer regarding these issues. In some cases, these rebates and deductions are not available unless a certified contractor installs your PV system. You'll also need to secure building and electrical permits before you can begin installing a system. Every geographical area is subject to federal, state, local, and village or town requirements that must be met or the entire project can be jeopardized.

When you consider costs of installing the PV system, you should keep these issues in mind. Permits cost money, as does hiring a professional installer. These costs should be included in your budget.

Warranties and Replacement Costs

Warranties are very important when determining the true cost of a solar project. Most solar panels are covered by warranties for a period of 20 to

30 years, and the amount of power they will produce is guaranteed during that time. Solar panels are the most expensive part of your PV system.

Low-quality panels can be used, and although they may cost less at first, they will not provide the consistent power you need. Some rebates and incentives may be contingent upon a certain quality of panel. Know and understand the warranties of all of the components in your system.

Also consider who is responsible for the replacement of bad panels or parts. Be sure to clarify who will install any required replacements and at what cost. Some manufacturers will provide service technicians to replace any faulty parts. Installers sometimes are reimbursed by the manufacturer for these issues, and some may offer you a separate service contract.

Buy your panels and installation from companies with a good and long history. New solar panels from China might be cheap, but will the manufacturer be in business two years from now? You get what you pay for.

Payback Time

You can determine the amount of energy your PV system will create during its useful life. Here's an example of how to do this:

A 10 kW system operates at average efficiency, 77 percent, for 6 hours each day, year round: this average factors in system efficiencies, number of sunny days, and other factors discussed in this chapter.

1. Start by figuring out the daily output of a 10 kW system:
   10,000 watts × 0.77 × 6 = 46,200 watts per day
2. Then figure out the monthly output:
   46,200 watts × 30 days = 1,386,000 watts/month
3. Then yearly:
   46,200 watts × 365 = 16,863,000 watts/year
4. Then factor in the following:
   - The degradation of the PV system: 0.25 or 0.5 percent each year for 30 years (numbers depend on manufacturer's specifications)
   - Increasing electricity prices

This will provide the total electricity your system will produce over its lifetime. Now you can calculate how much money you will save compared to typical electricity prices. In most cases, your payback time will be 10 years or less.

> The Department of Energy offers an excellent synopsis of the potential of solar energy. See http://apps1.eere.energy.gov/solar/cfm/faqs/third_level.cfm/name=photovoltaics/cat=The%20Basics.

Manual Disconnect Switch

If you are installing a PV system to provide power in case of a power outage, ask your PV installed or local electrician to add a manual or automatic disconnect switch, as shown here.

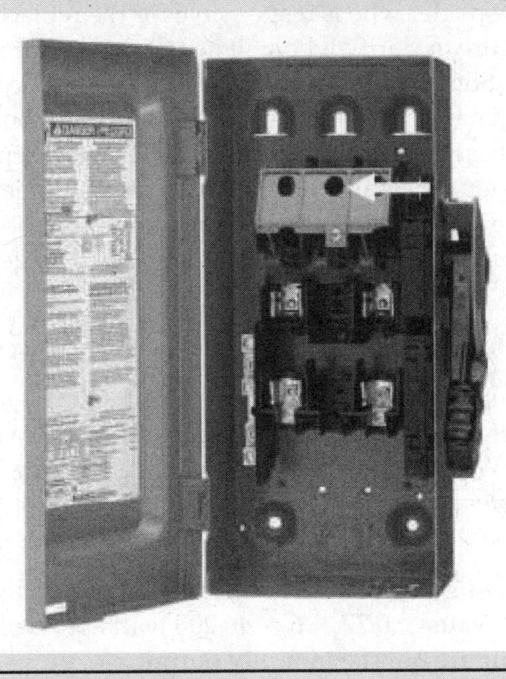

Manual disconnect switch
http://www.cdc.gov/niosh/face/images/06MA043b.jpg

This switch will allow you to disconnect from the grid and use the electricity that you are producing onsite. A supplemental power source, such as a backup battery, is required for this type of system. Hybrid systems can also be implemented with the addition of a generator, fuel cell, batteries, or other power source that can be supplemented with power from the grid.

CHAPTER 8
Solar Projects You Can Use Today

You can use solar power for small applications that can have a large impact on your energy consumption. The ideas covered in this chapter can save you energy and money and can help you reduce your carbon footprint. An array of solar and solar-powered products are available for many applications.

Solar-Powered Swimming Pool

Many places around the world, such as the American Southwest, Australia, and Southern Europe, can be extremely hot and dry, and they receive significant amounts of sunlight and little rain and cloud cover. Public and private pools are practically a necessity for the people who live in or visit these places. Swimming pools, however, consume large amounts of energy—but a pool need not be an environmental nightmare.

A solar pool pump (Figure 8-1) can be used and powered by a solar array that is grid-connected or independent (Figure 8-2). In addition, the pump can be a 12, 24, or any voltage complementary to the solar array to allow for more efficient use of the electricity produced onsite.

Heating is a large expense associated with a swimming pool. Consider the enormous amount of energy that is required to heat the average 80-gallon home hot water heater. Imagine the amount of energy required to heat your 50,000-gallon pool! Heating a pool, however, can be accomplished using solar energy.

A solar pool cover (Figure 8-3) resembles a large piece of bubble wrap. The solar cover floats on top of the water and traps the solar energy in the pool—similar to how global warming heats the Earth but on a micro scale.

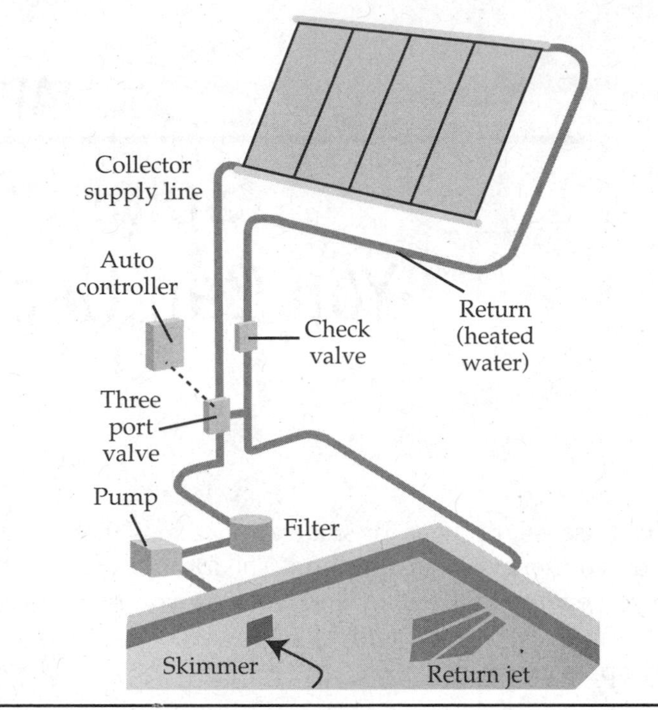

FIGURE 8-1 Solar pool pump
http://solarsaves.ca/images/pool_diagram.jpg

FIGURE 8-2 Solar panel array
http://vulcan.wr.usgs.gov/Imgs/Jpg/MSH/MSH07/MSH07_solar_panels_at_spine_5_site_05-25-07_med.jpg

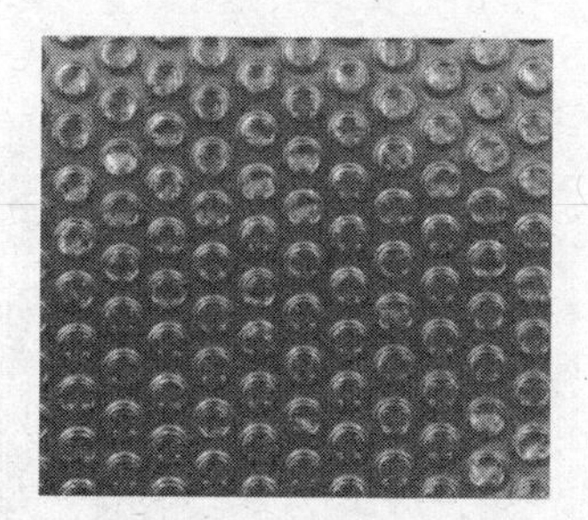

FIGURE 8-3 Solar pool cover heater

Here, your pool water is the Earth and the solar cover is the atmosphere holding in the heat from the sun. This type of solar cover can be deployed manually or electrically and stored in a housing transparent to the owner.

Solar water heaters can also be used (Figure 8-4) to heat water in large swimming pools.

Using a solar pool heater instead of a natural gas or electric-powered pool heater (Figure 8-5) can dramatically reduce your energy use and your electric bill.

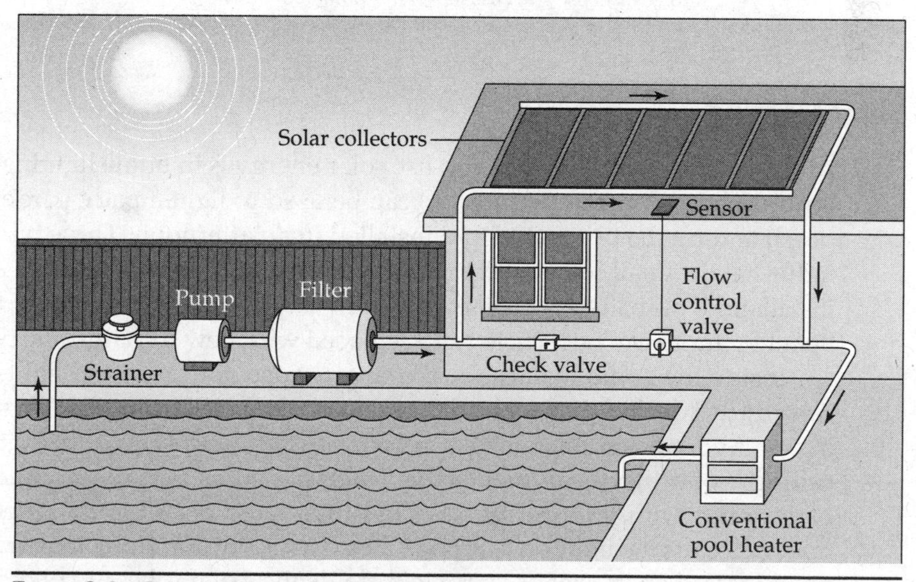

FIGURE 8-4 Solar hot water heater for a pool
http://www.energysavers.gov/images/pool.gif

FIGURE 8-5 Natural gas pool heater
http://www.cpsc.gov/cpscpub/prerel/prhtml03/03174.jpg

Solar Lighting

Another fantastic opportunity to use solar energy is in home lighting. Exterior lighting fixtures (Figure 8-6) can be used to light a path, garden, or porch and can be purchased and installed in an afternoon. These types of lights have a small solar collector and rechargeable batteries built in. After installation, the batteries may require a few days of sunlight to charge fully. Spent batteries can be recycled and replaced with new rechargeable types.

Inside the home, you can make smarter use of the natural light. You can adjust your workspaces to capture more lighting from windows and doors. And you can install solar tubes, or light tubes (Figure 8-7), in the ceiling. Solar tubes can be installed in small areas and rooms, such as bathrooms, with no exterior access. A tube lined with highly reflective material leads the light rays through a building, starting from an entrance-point located on its roof or one of its outer walls. Solar tubes are photo and thermally efficient, cost-effective, and easy to install.

FIGURE 8-6 Exterior lighting
http://www.airporttech.tc.faa.gov/Pics/DSC03584.jpg

FIGURE 8-7 Solar tube
http://en.wikipedia.org/wiki/Solar_tube

Another, more flexible, lighting option is fiber-optic lighting (Figure 8-8). Fibers of ultra-thin glass can be used to bring natural sunlight anywhere in your home. In fiber-optic lighting, a solar light collector (Figure 8-9) is placed in an area of direct sunlight. The collector, which looks like a parabolic dish, collects and focuses sunlight into fiber-optic cables, which carry the sunlight to the ends of the cables and into the room through a fixture (Figure 8-10).

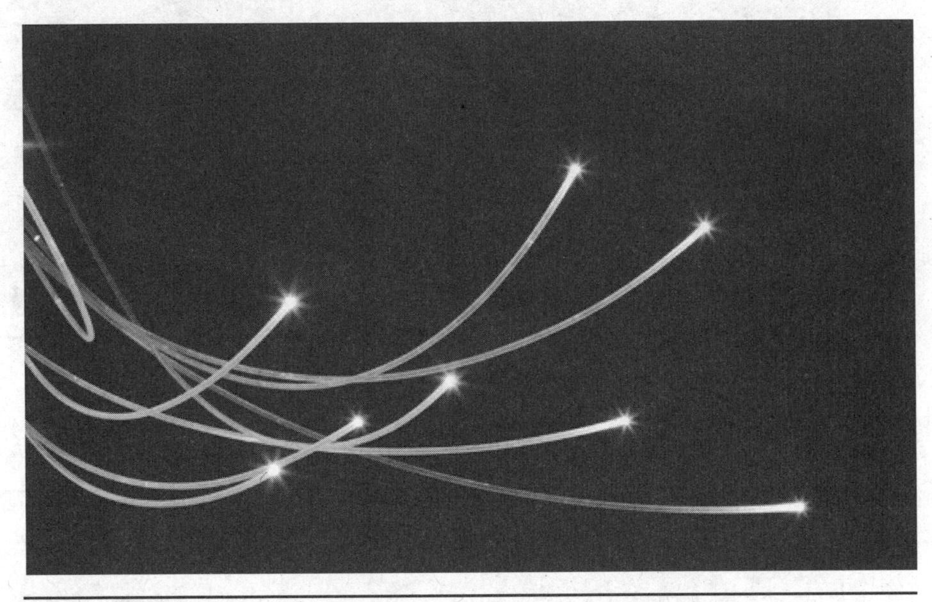

FIGURE 8-8 Fiber-optic lighting
http://www.ornl.gov/info/ornlreview/v39_3_06/images/a19_p10_hybrid_light.jpg

Solar Pet Home

In cooler climates, you can keep your pet comfortable by building a solar doghouse. See http://www.ehow.com/how_2136148_build-solar-dog-house.html for information.

Flexible Solar-Powered Gear

For fun, you can purchase an electric bicycle (Figure 8-11) that you can ride until you get tired, and then power your way all the way home. Electric scooters, cars, and carts are also available, and you can purchase portable flexible solar panels (Figure 8-12) to recharge your vehicle's batteries anywhere the sun is shining. Almost any device that uses electric-

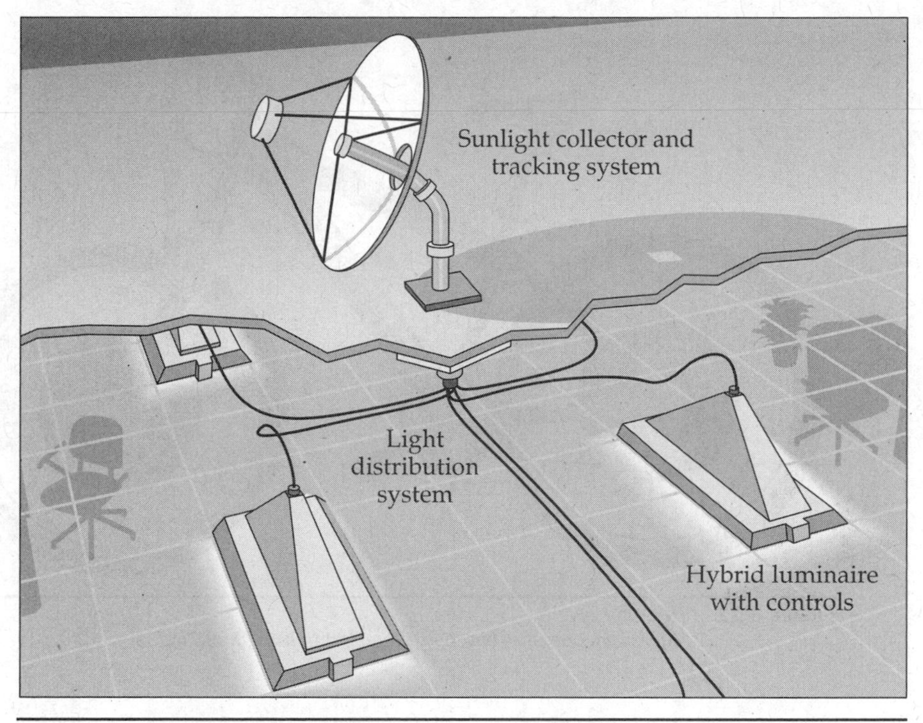

FIGURE 8-9 Fiber-optic collector
http://www.ornl.gov/info/ornlreview/v33_1_00/p22.jpg

FIGURE 8-10 Fiber-optic light fixture
https://www.llnl.gov/str/JulAug08/images/sensor2.jpg

Figure 8-11 Electric bicycle
http://www.cpsc.gov/cpscpub/prerel/prhtml05/05593.jpg

Figure 8-12 Flexible solar
http://www.sti.nasa.gov/tto/Spinoff2006/images/p066-079_img_13.jpg

ity or batteries can use portable solar. You can take it with you, roll it out, and have the comforts of home anywhere, any time the sun is shining.

You can place thin-film solar on a window sill and use it to recharge batteries. This type of solar is inexpensive, with many applications costing less than $100. Some universal 3-, 6-, 9-, 12-, and 24-volt chargers with multiple small applications can be purchased for less than $10.

Solar Oven

You can even cook your food with solar energy (Figure 8-13).

Solar ovens are being used successfully by households in economically disadvantaged countries such as Africa (Figure 8-14), where other alternatives are limited or do not exist.

Solar ovens can be purchased at a reasonable price, or you can make your own from cardboard and reflective material. Try it yourself using cardboard, tape, and aluminum foil. You will be surprised how effective the stove will be.

FIGURE 8-13 Solar oven
http://nihrecord.od.nih.gov/newsletters/2008/04_04_2008/images/story4Pic1.jpg

FIGURE 8-14 Solar oven in Africa
http://earthobservatory.nasa.gov/Features/RenewableEnergy/Images/senegal_solar_oven.jpg

Electric Lawnmower

A traditional lawnmower requires gasoline, oil, a spark plug, and an air filter. Electric lawnmowers are a better, cleaner option (Figure 8-15), and they can be charged by solar cells.

Electric-powered lawnmowers have always been available, but older electric motors required that a cord be plugged into an outlet, which was cumbersome and difficult to use. Today, electric lawnmowers are battery powered. The latest mowers come with two large and powerful batteries, so you don't need to plug your mower into an electric outlet. To charge the batteries, you can place a small solar charger on your shed and connect the batteries into the charger. The average homeowner mows the lawn once a week. During the week, the batteries can be continuously charged by the solar charger.

The cost of the average solar-charged electric mower is comparable to a quality gas mower. You can also find battery-powered lawn tools such as hedge trimmers, chainsaws, edgers, and more.

FIGURE 8-15 Electric lawnmower
http://www.cpsc.gov/cpscpub/prerel/prhtml02/02254b.jpg

Solar Carport

If you need a carport, you can build it with solar (Figure 8-16) on top. The solar carport can be used to power your home, outdoor projects, or your electric car. If funding is an issue, buy a single panel at first and continue to add panels over the years.

FIGURE 8-16 Solar carport
http://www.esrl.noaa.gov/gmd/obop/smo/images/Groof2.jpg

CHAPTER 9

The Real Costs of Energy Consumption

Most people who seek a photovoltaic (PV) solar system do not ask how much carbon dioxide (CO_2) their system will reduce over other energy-producing systems; instead, most ask how much money the PV system will save over conventional systems. According to Energy Information Agency (EIA) data from 2007, in most households, the following is true:

- The average monthly electric bill was approximately $100.
- The average kilowatt hours (kWh) used was 936 kWh.
- The average price paid per kWh was 11 cents.
- Households in high population density areas such as New York, Los Angeles, Toronto, and Tokyo pay significantly higher prices for electricity than households in smaller communities.
- During the next 30 years, with average inflation and very moderate increases in fuel costs, you will spend a minimum of $50,000 for electricity.

Global Warming

Another huge price we pay for consuming energy comes in the form of global warming, which isn't measured in dollars and cents, but in common sense. That is, if we cannot control global warming, we will all pay globally and in ways that are far more important than money.

Back in 1938, British amateur meteorologist Guy Stewart Callendar theorized that rising amounts of CO_2 in the Earth's atmosphere would contribute to an increase in global temperatures. This eventually became known as the Callendar effect.

In the 1950s, the U.S. government produced several educational films about the increasing CO_2 in the atmosphere and its effect on the Earth's temperatures. The film's producers actually proposed that global warming would be a good thing, allowing the Earth's permanent frozen areas to thaw and become suitable land for agriculture. Since then, the issue has crept into public consciousness; today, global warming is viewed as a seriously growing crisis (Figure 9-1).

We cannot afford to ignore this problem. We must address the issue because it will cause radical changes to the Earth's environment and every living creature inhabiting the Earth.

CO_2, You, and the World

Let's begin with what we cannot change. Humans and animals exhale CO_2 with every breath. According to various sources, on average, a human produces approximately 450 liters by volume, or 900 grams by weight, of CO_2 every day. How does this compare to nonorganic sources of CO_2? According to the U.S. Environmental Protection Agency, burning one gallon of gasoline produces 2421 grams of CO_2. Burning a gallon of diesel fuel produces 2778 grams of CO_2.

The gas or charcoal you use in your backyard barbecue releases CO_2. The gasoline you burn in your lawnmower, once combusted, releases CO_2.

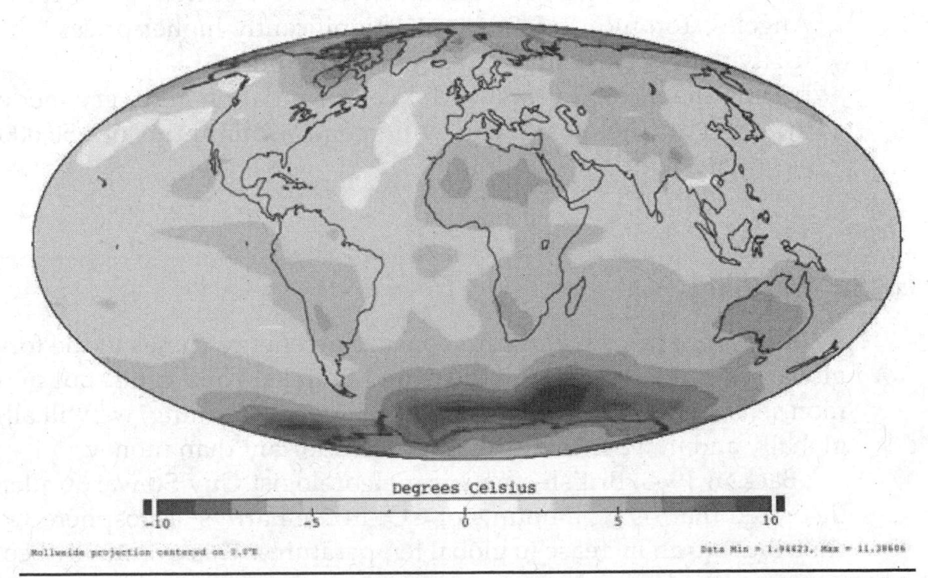

FIGURE 9-1 Surface air temperature (global warming) predictions from 1960 to 2060
http://www.nasa.gov/images/content/105582main_GlobalWarming_2060_lg.jpg

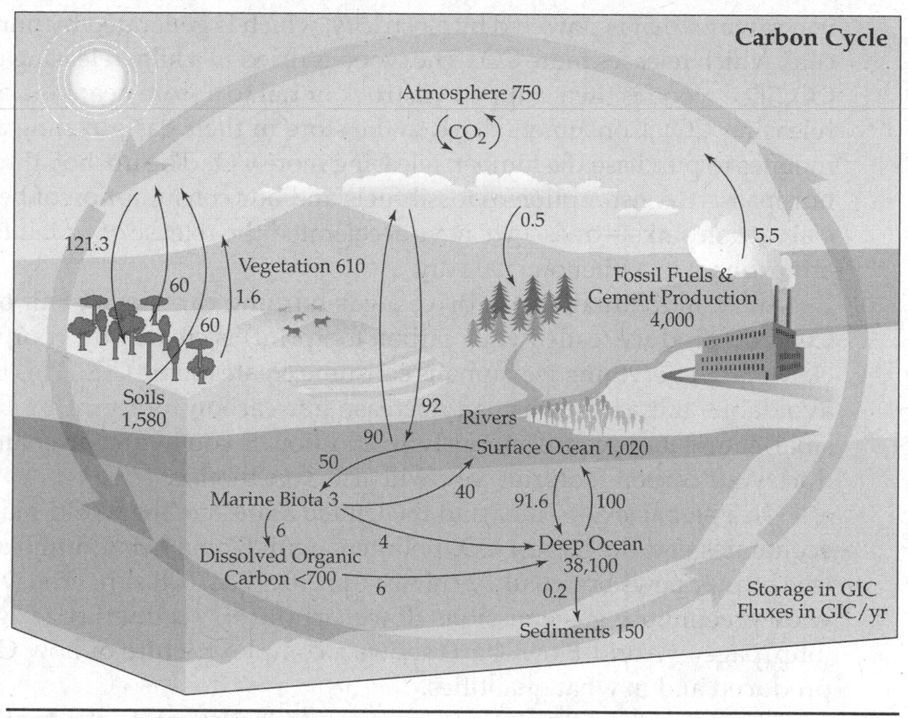

FIGURE 9-2 The carbon cycle
http://www.nasa.gov/centers/langley/images/content/174212main_rn_berrien2.jpg

And even as your grass clippings or tree trimmings degrade, they release CO_2 through a process called the carbon cycle (Figure 9-2), a biogeochemical cycle by which carbon is released into the Earth's atmosphere.

A tree offers a great example of how the carbon cycle works in the natural world. A tree can grow and thrive for hundreds of years. Throughout its lifetime, the tree consumes nutrients from the soil and carbon and oxygen from the air; it then releases water and oxygen into the atmosphere (which benefits us, of course, since we breathe oxygen and drink water). The tree (and, for that matter, the forest in which it is growing) is considered a "carbon sink" that stores carbon in living systems. When a tree dies, it begins to decay and returns some of the carbon to the atmosphere, but it does this very slowly.

Now think about what happens when we humans step into the picture. Humans plant forests to produce wood quickly to harvest for paper, lumber, and other products. Suppose a 500-year-old tree is cut down to use for lumber. Loggers use a chainsaw to cut down the tree; the chainsaw uses gasoline, a fossil fuel that produces CO_2. The logs are shipped to a mill in a diesel-burning truck, releasing more CO_2. The wood is milled in a sawmill that is powered by electricity, which is generated by burning coal, which releases more CO_2. The wood is dried in a kiln, releasing more CO_2. The wood is then shipped via truck or rail to a store near you, again releasing CO_2. Consumers drive to the store in their gas-guzzling automobiles to purchase the lumber, releasing more CO_2. The upshot: through our massive consumption of fossil fuels and our consumption of beneficial carbon sinks—trees—we have accelerated the release of CO_2 into the atmosphere in a phenomenal way.

As an individual, if you drive a gas-burning car, almost half of the CO_2 you produce (called your carbon footprint) is created by your vehicle. In fact, everything we humans consume creates some CO_2. This is unavoidable, but each of us can decrease our carbon footprint by living smarter and more conscientiously. In addition, if you live smarter and reduce your carbon footprint, you will also save money.

On a global level, China and the United States are the world's largest economies and the largest CO_2 polluters, with China being number one (see http://news.bbc.co.uk/2/hi/asia-pacific/7347638.stm). The worldwide greenhouse gas emissions flowchart (http://cait.wri.org/figures.php?page=/World-FlowChart) shows a complex picture of how CO_2 is produced and in what quantities.

The Kyoto Protocol (http://unfccc.int/2860.php), first adopted in 1997, is an international treaty aimed at reducing and stabilizing greenhouse gas emissions on a global scale. As of this writing, the United States has neither ratified nor resigned from the protocol; most other nations have signed and ratified the agreement and are attempting to reduce their CO_2 emissions accordingly.

Reduce Your Carbon Footprint

Truth is, governments and treaties alone cannot solve global warming. Corporations cannot solve global warming. The only group that can solve this crisis is you, as an individual.

If every individual today chose to use solar and other nonpolluting energy sources, we could start slowing global warming and begin to work toward controlling our CO_2 emissions. The first step is to become educated. The second is to take immediate action. You can start by determining your carbon footprint using a calculator such as the one found at www.nature.org/initiatives/climatechange/calculator/.

The calculator asks for information about your home, automobile use, appliance use, and other factors to determine your carbon footprint. You can use this information to create a plan of action to address the largest problems in your home.

The carbon footprint for a two-person household in the United States is currently 53 tons of CO_2 per year. On average, a human produces 4 tons of CO_2 each year, and each North American generates approximately 20 tons each year. Reducing your individual CO_2 footprint can be as easy as replacing or supplementing your traditional fossil-fuel use with solar energy; reducing your consumption of food sources that rely on heavy use of fossil fuels for production, such as meat; using Energy Star–qualified appliances; replacing incandescent light bulbs with energy-saving fluorescent models; insulating your home; voting for candidates who support sane environmental solutions; asking your utility company to use energy supplied by renewable resources such as solar and wind power; and installing a solar energy system in your home.

To blame any one person or nation for global warming is unfair. We are all responsible for global warming. Every individual needs to begin to make the necessary changes to benefit the Earth. As we begin to help the Earth, we help ourselves.

CHAPTER 10

How and When to Hire a Contractor

If you have decided that installing your own home solar project is beyond your physical or technical abilities, or if you hope to recoup costs by getting rebates or other incentives for installing a system, you can hire a contractor to do the work for you.

Contracting the work, however, is only the beginning. There is much more to contracting your home remodel than just hiring a service provider. Developing a superior and successful home remodel to make use of solar energy is a function of your planning, not just your contractor's abilities. Your builder or designer works to bring your ideas to life, so it is paramount that you understand what you want to accomplish, and that you plan every detail before the work begins. You can then convey your vision to your contractor.

Here's a quick checklist of issues that are important to consider:

- Interview several contractors.
- Check each contractor's references, memberships, insurance, and licenses. (Most governments require that a contractor obtain certification for solar installations.)
- Know what you want, and be specific in explaining your needs and ideas.
- Understand what to expect from the first meeting with the contractor.
- Do not be afraid to ask for more meetings throughout the project timeline.
- Understand what you should expect from a contractor.
- Do not settle for anything a contractor wants to sell you that does not contribute to your goals.
- Know your budget and compare costs before making a financial commitment.

- Make sure your contractor knows that you are in charge of the project.
- Do not change your plans unless you prefer (and can afford) an improved plan over the original.
- Plan the project in its entirety, including life disruptions and time overruns. Unless you are leaving your home during the renovation, personal and professional issues may interfere with the project timeline; and the project may interfere with your personal and professional timelines.
- Personally approve all contractors or service providers, including subcontractors.
- Know the project team: architect, designer, or other experts working on the project.
- Make sure that all service providers understand your plans, and make sure you understand theirs.
- Manage the project and check on the progress frequently.
- Do not be afraid to stop the progress or fire a contractor if the work is not progressing as it should or if the work is not meeting your expectations.

Deciding when and how to renovate your home to make use of solar energy involves planning, organizing, and sharing your plans with your family. Include everyone who will benefit from the renovation in the decision-making process.

Planning Your Solar Renovation

Proper planning should identify every detail, no matter how small or seemingly insignificant. Knowing what you want will allow for better communication with your contractor during the project.

As you create your plan, use the Internet to gather a bounty of valuable information that can include complete remodeling plan details, prices, and contractor information. Detailed product information of all types is available. You can find lots of information about passive and PV solar projects; PV panel types, brands, colors, and styles; and the prices of the units and installation.

You can start by creating the plan in reverse. Start by imagining how the finished project will look, and work your way backward to determine what needs to be done to create this image.

If you are remodeling a single room, for example, understand how the room will be used when it is complete. Consider the appearance of the room as well as the function of the space. Think about how and where sunlight will enter the room, how it will be distributed, how it will look as it shines on the walls, fixtures, furniture, and other objects in the room. Consider the seasonal light and sun angles and how they will affect the sunlight entering in the room.

Hire an Architect

Hiring an architect to design the project can cost you a few thousand dollars, but the benefits of using an architect are well worth the cost.

You can draw your own detailed plans that include structural elements, colors, furniture, entrances, windows, and doors, and show this to an architect at the beginning of the project. The architect will then use multiple communication tools to show and describe your project, including structural elements and materials used. The architect's plan will be clear, concise, and detailed. Hiring a professional architect also means that the plans will be checked by a licensed architect and possibly an engineer to ensure that they consider applicable laws and code restrictions. You can find an architect in your area by consulting the American Institute of Architects (AIA) website at http://www.aia.org.

Know Your Budget

Know your budget and remember that most projects go over budget by at least 20 percent due to unforeseen variables. Be sure to include extra funds in your budget to accommodate overruns; your contractor should be able to predict such issues in his or her cost estimates, but you should know how he or she comes up with these costs.

If you cannot afford the project, change your plans, not your budget. Do not attempt to deceive yourself and assume you will be able to reduce costs as your project progresses. Review your plans with someone you trust, including others who will be using the space, to be sure that your plans are reasonable.

You can negotiate with contractors—some offer variable costs so you can get a better price. The time to negotiate with your contractor is before you sign a contract and the work begins. If you negotiate for less cost after the work has begun, your contractor may be forced to cut corners.

Renegotiation because of unforeseen difficulties is a separate issue, and your contractor will understand these situations. Even expert contractors sometimes run into situations that they have never before encountered—such as opening up a wall and finding that the previous homeowner or builder did work that was not up to code or that was inadequate to support the changes you want to make. Reconfiguration of the plan is common and acceptable under such circumstances. But remember that plan changes always equal cost changes.

You can make changes to the original plan during the renovation, but such "change orders" always equate to more money. If you deem such changes worthwhile, and you can truly afford the extra costs, you can ask your contractor to make the changes. If you have planned properly, however, this most likely will not occur.

Choose the Right Contractor

Ultimately, you will be in charge of your home remodel, not your contractor. I cannot emphasize this point enough. Your goal is to work with a successful remodeling team that includes an architect, a contractor, and subcontractors.

The type of contractor you need depends on the size, scope, and complexity of your project. If you are merely installing insulation, a local handyman might fit your needs. However, if you want tax credits or rebates from the work done, you'll need to hire a professionally certified installer to do the work.

Most governments require that a certified solar installer is the only person who can install your solar panels to be eligible for the government rebates and credits. In addition, although your electrician may be able to install PV solar panels, a solar installer knows how to orient the panels, how much power they actually produce, and how to optimize your system.

Building an addition to your home will require a licensed contractor. Most contractors work with their own preferred subcontractors for plumbing and electrical work, but you can request that he or she use subcontractors that you prefer. Be sure that you ask what is possible before you sign the contract.

Installing new energy-efficient lighting may require the work of an electrician, but it may also require drywall and paint after the lighting has been installed. When choosing a subcontractor, attempt to choose a trade-specific subcontractor who is directly qualified to do the work required for your project. Always use a licensed, trade-specific contractor and subcontractors.

Many companies and contractors claim to do "environmentally friendly" or "green" work, but much of this is lip service only. Some companies using the "green" title are actually doing the same old contracting business as usual. Do your homework and know what you are purchasing before you buy it.

And don't forget to ask neighbors, friends, and family for contractor referrals:

- What was your overall experience with the contractor?
- How satisfied were you with the contractor's quality, professionalism, and the final result?
- How easy was the contractor to work with?
- How or were any subcontractors used?
- Would you use this contractor again?
- Was the remodeler available to answer question during the remodel?
- Did the contractor keep information about the project and its status?
- Did you have to change plans and was the contractor easy to work with during this process?

This is usually the best way to find a competent contractor. But remember that even though your friend may have had a good experience with a contractor, it doesn't mean that the contractor is licensed and insured for the type of work you need to do, and that you will necessarily be satisfied with the work the contractor does and how he or she does it. You still need to do your own homework. Ask your friends and family about bad experiences, too, because this can help you avoid choosing the wrong contractor.

In addition to asking for references when investigating a prospective contractor, do the following:

- Check with the Better Business Bureau (www.bbb.org) to inquire about unresolved complaints, numerous complaints, and fines; these are all indicators of a troubled contractor.
- Check with private building associations.
- Ask the National Association of the Remodeling Industry (NARI) for a private review of your contractor (www.nari.org).
- Ask about the remodeler's history.
- Examine the contractor's success for completion, schedule, cleanliness, and work habits.

Interviewing Contractors

Any prospective contractor should be friendly, trustworthy, and fair in price. The best contractors usually cost a little more but are usually worth the price because they pay attention to detail and use quality workers (and offer benefits), subcontractors, and materials.

Although you don't need to know how to accomplish your dream (that's the contractor's job), you must know what you want and be able to convey that information to the contractor in detail. Begin the interview by describing your dream or vision of the project. Present your plans, drawings, and specifications, including particular products you'd like to include. Providing the contractor with great detail increases the probability that he or she will perform to your expectations. Ask direct questions and expect direct answers.

If the contractor claims that he or she can perform the job, you can move to more serious questions about the contractor's company. Ask about insurance, years in business, licenses, complaints, and outstanding lawsuits. Do not be afraid to ask these questions, because quality contractors will offer you this information without hesitation. Let's review each of these issues.

Make sure your contractor is insured appropriately. If your project will cost $200,000.00, for example, and the contractor is insured for $50,000—well, you can do the math.

Ask how many years your contactor has been in business. Having a business for many years does not guarantee that your contractor is reputable. It only means that he or she has managed to stay in business. Do some research before the interview, check out the contractor's website (if one is available), and be ready to ask about the business.

Ask the contractor for a list of references, including former customers. The list should include names, addresses, and phone numbers of former customers.

Here's a personal example regarding a contractor I employed to replace my roof. I contacted the contractor based on a friend's referral. The owner of the company came to my home at my convenience. He gave me an exact price, not an estimate; he guaranteed his work and allowed me to choose from name-brand, quality materials. And he gave me a list of more than 100 former customers who lived within a half a mile of my home. This contractor also gave me a list of five people who were not immediately satisfied with his work. Three of the five people who had complaints did not like the color of their new roof, and they expected the roofer to buy new materials and pay for the labor to reshingle a brand new roof. The other two complaints were extremely complicated roofs with coordinated projects and other contractors. Minor leakages had develop from construction that occurred by a secondary contractor, after the roof had been completed. This roofer repaired these minor problems at no cost to the owners.

A contractor who gives you a list of the people who have filed complaints against his or her company is a confident contractor who will almost always be near flawless in performance. The contractor should be completely confident in every aspect of the job, including how his or her products or services relate to the other elements of your home.

During the first interview, also note the following:

- The contractor will have many questions regarding the scope of your project. This is a good thing if you expect the contractor to understand exactly what you need.
- He or she will ask if you have complete and detailed plans. If you've consulted an architect, now is the time to present the plans.
- The contractor should provide you with a list of the types of permits that will be required.
- Ask about products and materials that would work best for your project.
- Ask to see appropriate solar license and/or certifications.

Things to Notice During the Interview

You should watch for signs of trouble when interviewing a potential contractor. First, is he or she listening to you? If he or she does not take the time to understand your vision, how will the contractor be able to provide you with your dream? If your contractor tells you, "Do not worry, I'll take care of it," does he or she really mean that you are a nuisance and the contractor will do the job the way he or she wants to, regardless of your vision? Even if you simply "have a bad feeling about a contractor," politely say no thanks. Remember that the contractor will be working in your home every day.

Don't be afraid to ask the contractor to leave at any point in the estimation process or project. This is important. The further you are into the remodel, the more difficult it will be to replace a contractor. In addition, few contractors will want to attempt to assume someone else's unfinished project. Do not be afraid to take charge early if you need to do so.

Licensing

During the interview, make sure that the contractor's license is current. Be sure that his or her license allows for the type of work you need completed. Confirm that the license is appropriate for the geographical area in which the work will be done.

Ask about outstanding lawsuits. Contractors that have something to hide will try to avoid answering this question. Confident contractors will be honest and won't make excuses.

In addition, most PV installations require that a contractor have a solar installation license as well as a licensed electrician. In fact, most governments will allow tax credits and tax rebates to be released to you only if you use a licensed contractor.

Getting Estimates

Before you choose a contractor, try to get three or more contractors to offer written estimates for the work to be done, and let each contractor know that you are getting multiple estimates.

Try to spell out every detail for the project so that the estimates will be accurate. Know your products and understand what you are paying for. The more details you can specify, the better your contractor can understand, estimate, and create your vision. If you are unsure of some estimate details, ask the contractors for a second meeting. If you are unclear about any details, ask the contractor to explain the items until you are satisfied.

You may have the opportunity to negotiate a better price, which of course must occur before you sign any contracts. You can negotiate costs for the work, materials, labor, and time.

Ask how the work will be done. What hours will workers be in your home? If you want to expedite the project and want the contractor to work 12-hour days, including weekends, you must specify this before the contract is signed.

Will the contractor subcontract parts of the job? Has the contractor used these subcontractors in previous work? What can the contractor tell you about each? Do not assume the subcontractors will be licensed and insured. Ask for a list of subcontractors and ask to interview them. Ask each subcontractor if he or she is licensed for the type of work you need, if he or she is insured, and how long he or she has been performing this type of work.

Building Contracts

After you have interviewed and chosen a contractor, you'll be presented with a detailed plan, a timeline, and a written contract. At this point, you should have no concerns about whether your contractor is the right person to do the job. If you still have questions or concerns, you need to review these before you sign the contract, before you write the first check, and before work begins. A second interview or meeting to finalize details is common. Do not be afraid to call in the contractor for more meetings.

Negotiation of the contract should be as specific as possible. All terms and conditions need to be specified in writing and agreed to by both parties before work begins. A written contact will protect both the client and the contractor, and it will allow the work to progress smoothly. If you intend to do any of the work yourself or plan to hire additional assistance, such terms should be understood by your contractor and added to the contract.

The written contract should include the following:

- Contractor and homeowner's information
- License numbers
- Insurance information
- A detailed description of all work to be completed
- Start and completion date, and consequences if the dates are not met
- An itemized inventory of materials
- Specified quality and quantity of materials to be used
- Identified colors, brands, sizes
- Documented itemized inventory of all costs, including labor
- Created and agreed-upon schedule for payment
- A detailed plan for debris removal
- Guarantees and warranties for both the contractor and the material
- The right of refusal—ability to cancel the contract before work begins or material is purchased
- The right to cancel or renegotiate the contract if unforeseen problems arise
- A signature and date

During your final meeting with the contractor prior to signing the contract, both parties will sit together and edit the contract as needed. Most preprinted contracts have space where you can write in your specific requirements. Then both parties sign and date the contract, and both parties also receive a copy of the contract.

In most states, you have the right to cancel your contract within three days of the initial signing. Each state or country has its own rules and regulation. Be knowledgeable before you sign the contract. The contractor should inform you of the right to cancel. He or she should also explain the process of cancellation; if the remodeler does not inform you, ask what your rights are according to the contract.

Payments

After the contract is signed, payments may be disbursed. Most contractors follow the informal rule of thirds payment plan and will expect a check or cash for the first third of the amount estimated for the job before work begins. The next third will be due when the job has begun and is partially completed, and the last third will be due when the job is complete. If you have not worked with your contractor previously, you should attempt to pay a low down payment until the contractor has begun work. You should not pay the second or final third if the work is behind schedule. Payment should be made for phases completed, and not time intervals. Payments should not be made for incomplete or poor-quality work until you are satisfied with the work. All of these items should be identified clearly in the contract.

Most firms of significant size and expertise will not ask you for any money until the job is complete. These firms are confident in the materials they use and the workmanship they employ.

Finally, your contractor may want to be compensated for his or her time during the interview process, particularly if you have an extensive remodel and your contractor spends days helping you develop a plan. This is an acceptable payment.

Never provide the final payment or release until you are satisfied with the work completed. Verify that all subcontractors have been paid. If the subcontractors or supplies are not paid by the contractor, you may be liable. You can completely avoid this problem with proper planning. As a client, you can require a "Release-of-Lien" clause to the contract. Payments to the subcontractors can be placed in escrow and not disbursed until work is completed to your satisfaction and to the satisfaction of your general contractor.

Warranties of materials or workmanship should be in writing. Careful attention to detail will prevent any misunderstanding. Understand all terms and conditions of the warranties. Be familiar with the length of time, physical limitations, geographical locations, and your rights as the consumer of this product and/or service.

Permits

The contractor will obtain the building permits. Proper permitting is vital; failure to get the proper permits before work begins could cause significant delays or fines for noncompliance with the building codes. Building codes are established at all levels of government (such as the insulation requirements shown in Figure 10-1).

Building codes are implemented for your safety and your benefit. You may not understand or agree with the building codes in your area, but the permit process is an absolute must to ensure your safety and the safety of the building project. Plus, you can be fined without proper permits, and

					Ceiling						Basement	
Zone	Gas	Heat pump	Fuel oil	Electric furnace	Attic	Cathedral	Wall (A)	Floor	Crawl space (B)	Slab edge	Interior	Exterior
1	✔	✔	✔		R-49	R-38	R-18	R-25	R-19	R-8	R-11	R-10
1				✔	R-49	R-60	R-28	R-25	R-19	R-8	R-19	R-15
2	✔	✔	✔		R-49	R-38	R-18	R-25	R-19	R-8	R-11	R-10
2				✔	R-49	R-38	R-22	R-25	R-19	R-8	R-19	R-15
3	✔	✔	✔	✔	R-49	R-38	R-18	R-25	R-19	R-8	R-11	R-10
4	✔	✔	✔		R-38	R-38	R-13	R-13	R-19	R-4	R-11	R-4
4				✔	R-49	R-38	R-18	R-25	R-19	R-8	R-11	R-10
5	✔				R-38	R-30	R-13	R-11	R-13	R-4	R-11	R-4
5		✔	✔		R-38	R-38	R-13	R-13	R-19	R-4	R-11	R-4
5				✔	R-49	R-38	R-18	R-25	R-19	R-8	R-11	R-10
6	✔				R-22	R-22	R-11	R-11	R-11	(C)	R-11	R-4
6		✔	✔		R-38	R-30	R-13	R-11	R-13	R-4	R-11	R-4
6				✔	R-49	R-38	R-18	R-25	R-19	R-8	R-11	R-10

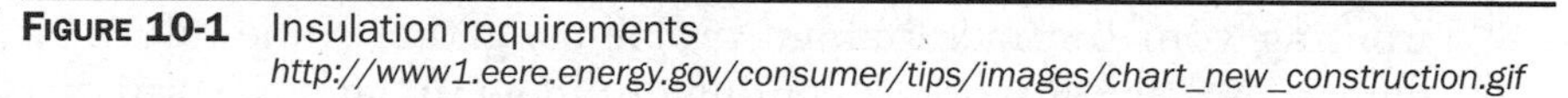

FIGURE 10-1 Insulation requirements
http://www1.eere.energy.gov/consumer/tips/images/chart_new_construction.gif

you can be asked to tear down any unpermitted construction. Finally, not securing building permits can cause problems when it comes time to sell your home. During the sale of your home, you will need to obtain the proper permits and possibility redo the work originally completed without the permits.

Zoning and permitting laws vary considerably from one jurisdiction to another. Be sure that your contractor is familiar with the process of obtaining the appropriate permits. Your contract should state that the work performed will be in accordance with the applicable building codes in your area. Work done by subcontractors, such as electrical, heating and cooling, and plumbing, also requirs permits.

Licensing and permitting are almost always approved for solar improvement projects, because most municipalities want to assist you in improving your home. Your home improvements theoretically better your neighborhood and therefore your tax base.

As the work is underway, your contractor will arrange for city inspectors to sign off on the work completed. The inspector will come to your home and verify that the work performed complies with local building code. If the contractor's work does not pass inspection, you cannot be held responsible financially for the corrections that need to be made. All issues and items should be negotiated with the contractor before work begins and this should be included in the written contract.

Permits are particularly important with green renovations and solar projects. The contractor and homeowner must adhere to strict rules to obtain tax exemptions, rebates, and credits. Failure to follow correct procedures will result in the forfeiture of all of these incentives. For most PV systems, this would mean a loss of tens of thousands of dollars.

In most building codes, a solar project is considered a home improvement, and although it will cause your local government to raise your property tax, most governments can provide a tax waiver. Local governments will allow individuals to make changes that are beneficial for the individual and community without incurring additional taxes. Confirm this with your installer or contractor.

What to Do If Things Go Bad

If problems occur, resolve the issue as soon as possible. Any other alternative will cost you significant amounts of time and money. Before confronting your contractor about problems, gather some information. If your contractor or workers continually show up late, be cordial; inquire as to why he or she is not acting as promised. Before reprimanding, understand that the contractor has a life beyond work; be understanding. If the contractor is showing up late because he is working on another job, however, you have a right to be upset and to attempt to correct the problem. Identify and deal with problems or conflict immediately. Address all issues before they become a significant cost or time concern.

What is important during any conflict is resolution with the contractor. Try to deal directly with the contractor to complete or repair any problems free or for a reasonable fee. Such negotiation is usually the least expensive and painful option for the homeowner. If you cannot speak to the contractor, you can file a complaint with the Better Business Bureau and your state consumer complaint board. These organizations will contact the contractor and attempt to act as an intermediary to resolve the issue.

Resources and channels for consumer complaints include the following:

- Better Business Bureau: www.bbb.org
- Federal Trade Commission: www.FTC.gov (FTC) or 877.ftc.help, or 877-382-4357
- National Association of the Remodeling Industry (NARI): www.nari.org, or 703-575-1100

- National Association of Home Builders (NAHB): www.nahb.org, or 202-822-0216
- National Association of Realtors: www.realtor.org, or 202-383-1000

When filing a complaint, you will begin with your local state office of consumer protection. If a fraud has occurred or legal action is necessary, you can contact you state attorney general's office and they will contact the remodeler directly. Most organizations offer online claims forms that make the process simple. When filing a complaint, state the facts and leave out emotions or examples. These organizations care only about the law.

Your final option, if you are seeking small retribution, is small claims court. The process is simple and straightforward, and a judge makes the decision immediately. If your remodel was expensive and you cannot resolve the issue in small claims court, you'll need to hire an attorney and take the contactor to court.

CHAPTER 11

Funding Your Solar Project

This is probably the best time in history to remodel and use energy-efficient systems and products in your home, because the available incentives and rebates are great in scope and size. This chapter offers suggestions for financial options to help fund a solar project or any energy-efficient upgrade in your home.

Your first tasks are to reduce energy usage, increase efficiency, and use or add passive solar. Then you can purchase a photovoltaic (PV) system for your home. Each of these projects can be at least partially funded by a variety of programs and tax rebates.

Savings

The money that you have accumulated can be put to work to make your energy dreams come true. Using your savings is the friendliest way to fund your project. There are no loan applications, interest rates, or paperwork. In addition, if you can pay in cash, many contractors or suppliers will offer you discounts.

And Loans

The most common options available to consumers who want to fund home projects are loans from many sources:

- Traditional loans
- Low-interest (typical for PV solar) loans

- Zero-interest loans
- Deferred loans
- Home equity loans
- Mortgage refinancing

If you cannot pay for the home improvement project without a loan, add a clause in the contract with your builder stating that the agreement is contingent upon financing. This will allow you to cancel the contract if it turns out you are not eligible for financing.

Bank Loans

Traditional loans from a bank, credit union, or other lender can be obtained provided you have a good credit score, a steady income, and some equity in your home. Banks, credit unions, and other lenders offer secured loans on the Internet. Internet sites such as Lending Tree gather multiple quotes for you after you fill out a brief, non-binding application. These Internet resources can provide viable lenders with competitive rates. You can compare rates and fees at your convenience, in the comfort of your own home.

Contractor Loans

Your contractor or home superstore may be willing to finance your home remodel. The provider will benefit from this arrangement in three ways:

- They will sell you a product.
- They will sell you a service.
- They will earn interest by financing your project.

You should secure this type of loan only if you understand and accept the terms of the loan. Many box stores or large remodel firms will finance your home remodel project for a limited time or a low introductory rate, which may end soon after the contract is signed. These businesses are not required to provide you with the best deal. Although these loans can be very competitive, they can also be very expensive.

Credit Cards

Credit cards are not long-term financial instruments. Credit card terms can be changed at any time, for any reason, or for no reason at all. You don't want to get stuck paying 33.9 percent interest for your home improvements. If you are using your credit cards to be paid in full at the end of the billing cycle, or prior to a loan approval, it's reasonable to finance your project temporarily with credit cards.

FHA and HUD Loans

For homeowners in the United States, the Federal Housing Administration (FHA) often sponsors these loans. (State or local government may have similar programs.) Your bank will be familiar with these loans and can help you apply for funding. The green revolution has created opportunities for some home improvements that may be financed with FHA loans.

Low-interest loans are available for energy improvements in your home, new energy-efficient construction, and for other purposes. Visit http://portal.hud.gov or http://www.hud.gov/ for details on the types of loans available and their restrictions.

Federal funding can come from a few different agencies, including Housing and Urban Development (HUD). Although HUD is a federal agency, it does offer programs that provide low-interest loans for homeowners in individual states. The web link http://www.hud.gov/local/ny/homeownership/homerepairs.cfm, for example, lists resources in the State of New York that are available for homeowners. Each state's program is listed with HUD at www.hud.gov. HUD is more than just about funding. At the HUD website, you will find resources to help you better your home and your community. You can find information about financial counseling, predatory lending assistance, safety requirements for buildings, links to agencies for assistance, and much more.

Another HUD program is "Funds for Handyman-Specials and Fixer-Uppers" located at http://www.hud.gov/offices/hsg/sfh/203k/sfh203kc.cfm. This is the HUD 203(k) program, and it is available for homeowners who purchase or refinance a property. The program allows the buyer to borrow the purchase price of the property plus the funding needed to rehabilitate the property. The loan is FHA insured and is provided through approved mortgage lenders in your area. You must possess a good credit score to apply. The loan will include the purchase price of the home, the cost of repairs, and a 10 to 20 percent contingency reserve for unforeseen remodeling issues. A list of HUD 203(k) lenders is available at HUD's website.

Another HUD program, the HOME Investment Partnerships Program, is the largest federal block grant to state and local governments designed exclusively to create affordable housing for low-income households. The program assists homeowners with the repair, rehabilitation, or reconstruction of owner-occupied units. The funding may be used for weatherization, emergency repair, or handicapped accessibility programs to bring a property up to code. For more details go to http://www.hud.gov/offices/cpd/affordablehousing/programs/home/.

Veterans Loans

A specialty loan that has been available to veterans is the Veterans Association Loan Program. This program has been around since after World War II, when veterans received help in purchasing their first homes. The United States Department of Veterans Affairs (the VA) website is at http://www.va.gov/. You can access the loan page directly at http://www.homeloans.va.gov/. The VA website includes all of the information needed to direct you through the lending process.

USDA Loans

The United States Department of Agriculture (USDA) Rural Development assistance program (http://www.rurdev.usda.gov/) provides assistance to select borrowers for repairs of existing homes. The loans are available in rural communities and small incorporated towns of up to 10,000 people, but some communities of between 10,000 to 20,000 people may be eligible.

Other Federal Loans

Two additional types of federal loan options also deserve mention. In deferred loans, the interest, or principal, or both can be deferred for a specified period of time. Most people are familiar with these types of loans, which are used primarily for education. These loans sometimes contain an option attached to government-sponsored loans, called "loan relief." These loans do not have to be repaid, provided the conditions of the loans were achieved. These loans are unique and all conditions must be met to qualify.

State-Funded Loans

On state and local levels, you can find many types of services to help promote and benefit you and your community. A good example of homeowner assistance is from the Connecticut Department of Social Services (DSS) at http://www.ct.gov/dss/site/default.asp. The DDS administers the Connecticut Weatherization Assistance Partnership with utility companies and local community action agencies. The program assists low-income families with incomes up to 200 percent of the federal poverty guidelines to reduce their energy bills by making their homes more energy-efficient.

Home Equity Loans

Home equity loan are excellent options for people who own their home. A home equity loan payment is added to your mortgage. The loan can be open or closed. In an open loan, you can borrow and repay capital as you require. A closed loan is similar to a mortgage: you borrow a fixed amount of funding and repay that funding on a prearranged schedule. Check with your lender and your accountant for tax considerations before you make a financial commitment.

Mortgage Refinancing

Your final loan option is to mortgage or refinance your current mortgage. If you own your home outright and do not have a current mortgage, you will have no difficulty refinancing your home. You can borrow as much as 80 percent of the current value of your home. If you currently have a mortgage, the option to borrow is the same, provided you have equity in your home. Here's how it works:

- Your home mortgage was originally $500,000
- You currently owe $100,000 on the mortgage
- Your bank says you can borrow 80 percent of the equity of the total value
- Equity available: $400,000
- You can borrow $320,000

So why can't you borrow 100 percent of the total value of your home? The mortgage lender/holder, usually a bank, requires 2 to 10 percent to resell a home if the home goes into default (you cannot pay your mortgage). In addition to these banking expenses, if the mortgaged real estate drops in value, the owner will have borrowed more money than the home is worth. If the borrower of the loan defaults, the bank will lose money.

The benefits of a new mortgage are numerous. You can choose the terms that best suit your needs; you can borrow the amount that you need; you can shop for the best lender, which lets you select the best rate and loan terms. You can "lock in" a fix rate that cannot be changed. Finally, your mortgage interest is tax-deductible. Using the equity in your home is often the best solution for home improvements.

Rebates, Tax Incentives, and Tax Credits

Following are some examples of other options that can help you finance a solar energy project. These programs are helping many people add solar power systems to their homes.

PV systems are being supported by many governments in Europe and in the United States, where rebates are available from the federal government. Financial benefits may be available from state and local governments in addition to the federal benefits in the following forms:

- Tax rebates
- Tax credits
- Projects that pay for themselves, such as solar, wind, or geothermal

Tax Rebates

Tax rebates are simplistic in design and allow a homeowner to make energy-efficient home improvements and get a rebate in the process. To take advantage of tax rebates, you must use products or services that qualify for a tax rebate. Application for tax rebates are prepared and submitted separately or completed with your income tax return. Tax rebates are granted to individuals regardless of tax status.

Tax Credits

Tax credits are different from tax rebates in that tax credits are applied to offset tax balances. Tax credits, like tax rebates, are designed to be simple. By installing eligible products and services to improve the energy-efficiency of your home, you can claim a tax credit. You are usually required to fill out paperwork and submit it before you can get a tax credit, so be sure you know the process in your area. Tax credits are applied when you file your annual tax return; be sure to mention these purchases to your tax accountant when applying for your refund.

Grants

Grants for energy products and services may be available. Begin by asking your local government if such funding exists at the local level. Local grant money is often provided to improve the energy efficiency of homes using windows, roofs, insulation, and so on. Your state government may also offer programs to which you can apply for a grant.

Federal grants are also available, and you can find plenty of options at http://www.grants.gov. Finding the grant that is perfect for your situation will require time and effort on your part. Researching and writing a proper application will be time consuming.

Your final option is private grants through non-profit organizations, foundations, and trusts. Much like government grant applications, the process is complicated and takes time. Use the Internet to look for grant sources.

Selling Power Back to the Grid

Suppose you place a solar power system on your roof that creates more electricity than you can use. You can sell that electricity back to the utility company or to other private buyers. This may not only fund your solar project, but can provide income for you once you have paid for the financing of the project.

CHAPTER 12

The Future of Solar Energy

Rising fossil fuel prices, increased energy requirements, improvements in materials science, and many other factors are making solar energy more valuable than ever. Solar energy offers many advantages now and in the future. The future of solar is bright.

If you've read the book this far, you now know that solar energy is more than just installing photovoltaic (PV) panels on your roof. And in the future, a new world of solar products will be available, including nano solar, bio-solar, and chemical solar, which are being created through a convergence of technologies that was not available in the past.

The future for solar energy as an industry is excellent, provided that governments continue to support the growth of solar energy. Renewable energy currently receives energy subsidies similar to those offered for coal and other fossil fuel production (http://tonto.eia.doe.gov/energy_in_brief/images/charts/share_of_subsidies_large.gif).

Thin-Film Solar

Thin-film solar (Figure 12-1) has been discussed throughout this book as an important new technology. Thin-film solar can be produced in large quantities at a rapid rate.

Thin-film solar has the most potential of any new solar technology. With a 19 percent efficiency rate shown in the laboratory, it produces less energy than PV panels in direct perpendicular sunlight, but thin-film produces more energy than PV panels in indirect sunlight and when the angle of the sunlight is not perfectly aligned.

FIGURE 12-1 Thin-film solar panels
http://www.nrel.gov/pv/thin_film/docs/springerville2003_half_a_mw_fs_cdte.jpg

Thin-film can be used on a roof, similar to other PV panels, and it can be formed and colored to conform to other shapes and sizes, such as roof shingles (Figure 12-2).

Small projects, such as solar chargers of any size, can be designed and built to meet the needs of any application. Converging technologies are creating new products in conjunction with thin-film solar.

You can find a directory of thin-film solar technologies and resources at http://peswiki.com/index.php/Directory:Thin_Film_Solar.

Micro Solar

A modified form of solar capture, micro solar (Figure 12-3) offers "glitter-sized" PV power. Micro solar cells can be applied to buildings, cars, and even clothing.

Micro solar may be able to produce the same amount of electricity as traditional solar cells, but using 100 times less silicon—the energy-producing and most expensive component in any PV system.

FIGURE 12-2 Thin-film roof shingles
http://www.nrel.gov/pv/images/photo_07157.jpg

FIGURE 12-3 Micro solar cells
http://www.sandia.gov/news/resources/news_releases/images/2009/pv_micro.jpg

Nano Solar

Nano solar cells are even smaller than micro solar; nano solar has the greatest promise because of a host of emerging nanotechnologies.

One of the greatest promises comes from the Idaho National Laboratories, which is working on a "nano-antenna" technology that could turn out to be 80 percent efficient. This film is so small it can be applied anywhere, and so efficient it could capture energy from moonlight. A brief but interesting article can lead you to more information about this type of nano solar: http://www.ecogeek.org/content/view/1329/.

Many companies are claiming that because of increased production, nano solar electricity will be more affordable than coal energy, currently the lowest priced fuel for energy production (http://techcrunch.com/2007/12/18/nanosolar-is-gunning-for-coal/). Several companies claim that they will be producing nano solar panels that will produce energy costing only $1 per watt of electricity. Such benefits could mean a nano solar revolution in the future.

Because solar panels can become up to 30 percent less effective when they accumulate dust, dirt, and bird droppings, researchers at Tel Aviv University have created a new nano-material that repels dust and water. The material could be applied as a sheer coating, creating self-washing windows and solar panels—Teflon for your solar panels. For more information, see http://www.inhabitat.com/2009/12/10/new-nano-material-paves-way-for-self-washing-solar-panels-and-windows/.

The U.S. Department of Energy discusses new developments in all energies on its website at http://apps1.eere.energy.gov/news/progress_alerts_archive.cfm.

Light-Spectrum Technologies

You'll recall that PV solar collects light from a small band of the light spectrum (Figure 12-4). Some new forms of thin-film solar focus on collecting more of the light spectrum to produce additional energy for the same square feet. Lawrence Berkeley National Laboratory, working with crystal-growing teams at Cornell University and Japan's Ritsumeikan University, are at the forefront of such technology: http://www.lbl.gov/Science-Articles/Archive/MSD-full-spectrum-solar-cell.html.

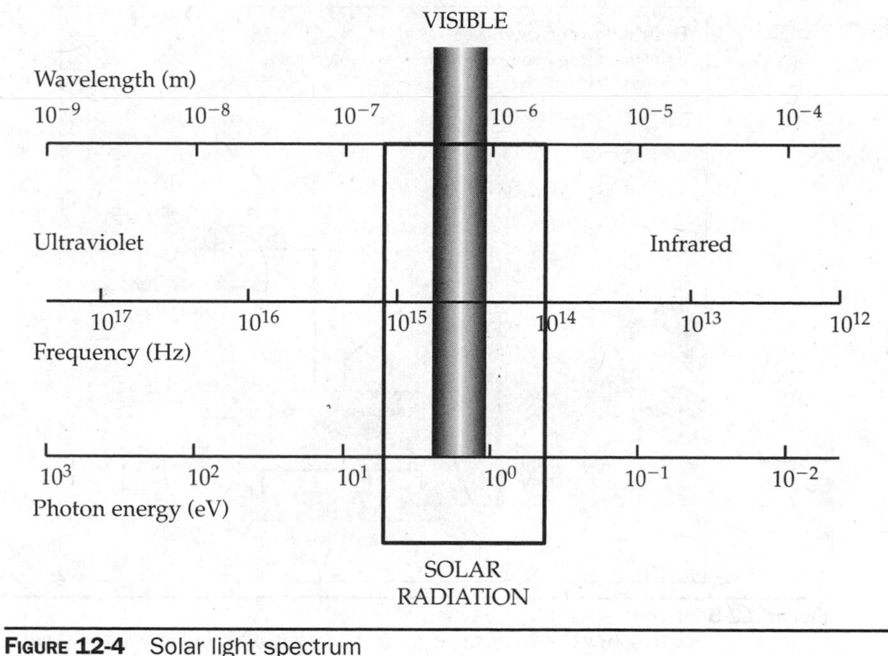

FIGURE 12-4 Solar light spectrum
http://www1.eere.energy.gov/solar/images/illust_radiation.gif

Hybrid Technologies

Hybrid solar technologies (Figure 12-5) involve the integration of multiple technologies that allow the homeowner access to off-the-grid electricity 24 hours a day, every day of the year.

In a hybrid solar energy system, PV panels are part of an integrated system. When the sun is not available, these systems use stored solar energy from grid electricity, hydrogen fuel cells, and batteries to disperse electricity.

Electrical Grid

The utility grid, while not the most efficient form of energy, does provide a good backup or a secondary system for your PV system. As long as you produce as much energy as you can through solar, you can tap into the grid for supplemental energy, as needed. The grid is like a big battery.

Hydrogen Fuel Cells

Hydrogen fuel cells can be used to store solar energy (Figure 12-6). At the moment, this is a very expensive option. In this system, the PV solar pan-

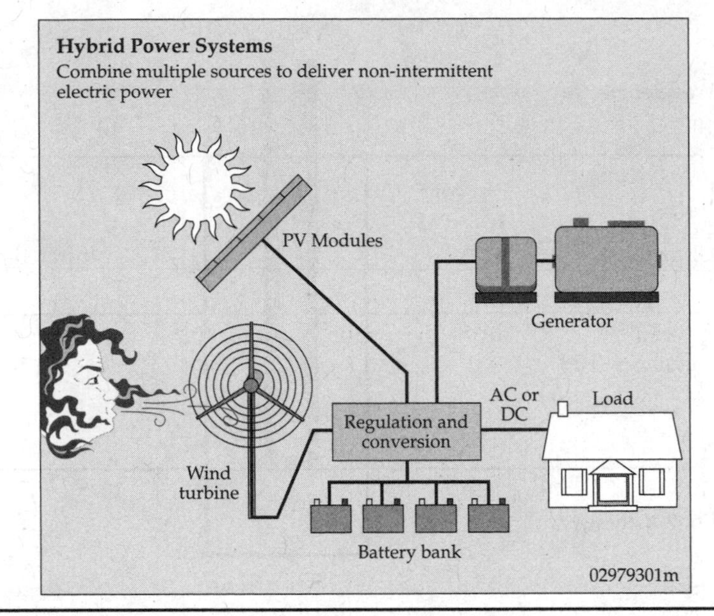

FIGURE 12-5 Hybrid solar technologies
http://www1.eere.energy.gov/buildings/residential/images/wind-powered_electric_systems_3.gif

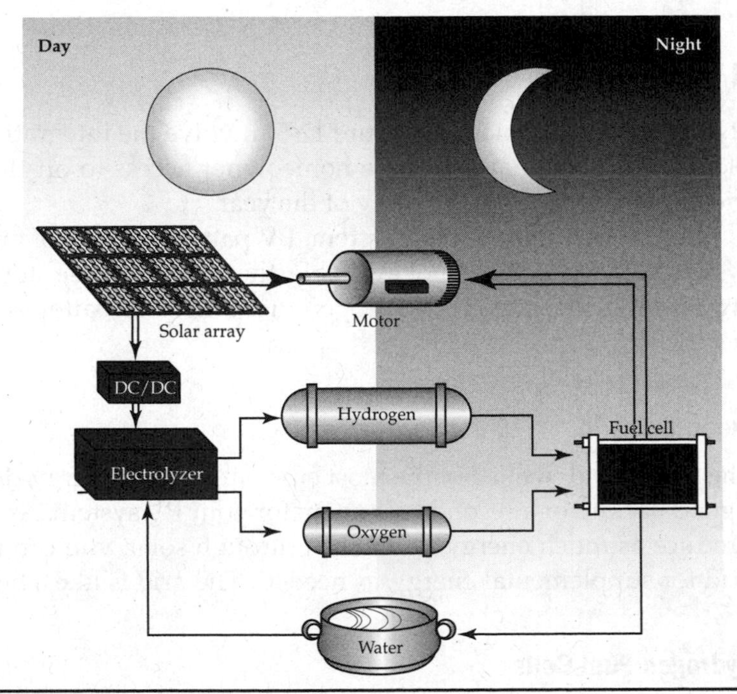

FIGURE 12-6 Hydrogen fuel cell system
http://www.nasa.gov/centers/dryden/images/content/107641main_Helios_diagram.jpg

els produce electricity, and the surplus electricity is converted into hydrogen and stored in fuel cells. When sunlight is unavailable, the stored hydrogen in converted back into electricity.

Battery Backup

The third viable option today is PV solar with a battery backup system. Figure 12-7 shows the Hubble telescope battery system that functions pretty much the same as any other PV and battery system would function.

When the sunlight is available, the electricity it produces charges the batteries. After the sun sets, the system uses the energy stored in the batteries. This type of system can be completely independent of the grid. The only limitation is the size of the PV system required to produce electricity and the size of the battery bank to store the potential energy. Even so, a similar system is affordable for many homeowners.

Storage Systems

Mechanical storage includes the mainspring in a watch, the water from a fast-flowing stream, or a dam (Figure 12-8); these are all forms of stored or potential energy.

FIGURE 12-7 PV solar with battery backup
http://www.nasa.gov/images/content/274993main_battery3.jpg

FIGURE 12-8 Hoover Dam
http://www.usbr.gov/lc/hooverdam/images/C45-20485-L.jpg

Thermal storage for homeowners can be as simple as opening the blinds in your home. The sunlight heats the floor or wall and then releases the energy as heat when the sun is no longer available. This, of course, is passive solar. It is immediate and delivered pollution-free and cost-free.

Thermal storage can also be used on an industrial scale (Figure 12-9). Industrial thermal storage concentrates solar energy, using water, oil, salt, or other media for heat transfer. Concentrating solar uses a fluid heated to greater than 1000°F. After the sun has set, the heat stored in the liquid is used to create steam and turn a turbine to produce electricity.

The Promise of the Future

The promise of hybrid solar and solar as a complete energy form is tremendous, from the smallest example (Figure 12-10) to the most complicated. Using solar, you can become electrically independent.

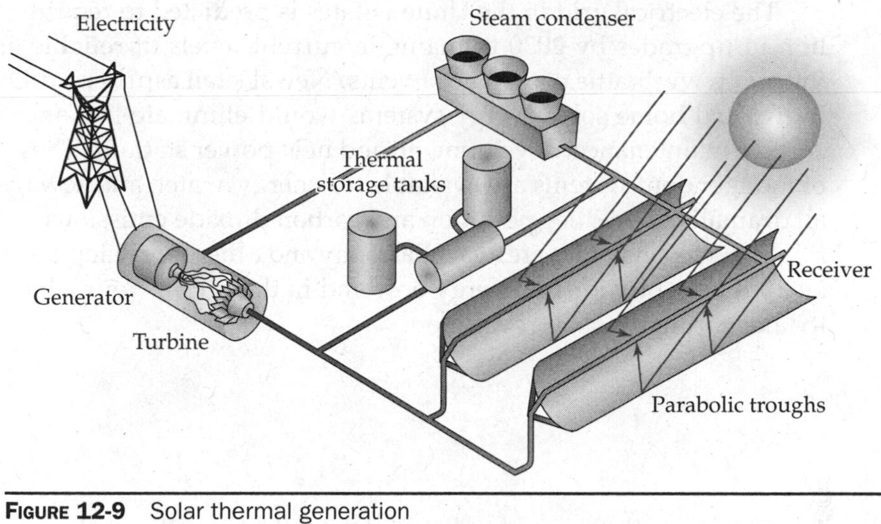

FIGURE 12-9 Solar thermal generation
http://www1.eere.energy.gov/solar/images/parabolic_troughs.jpg

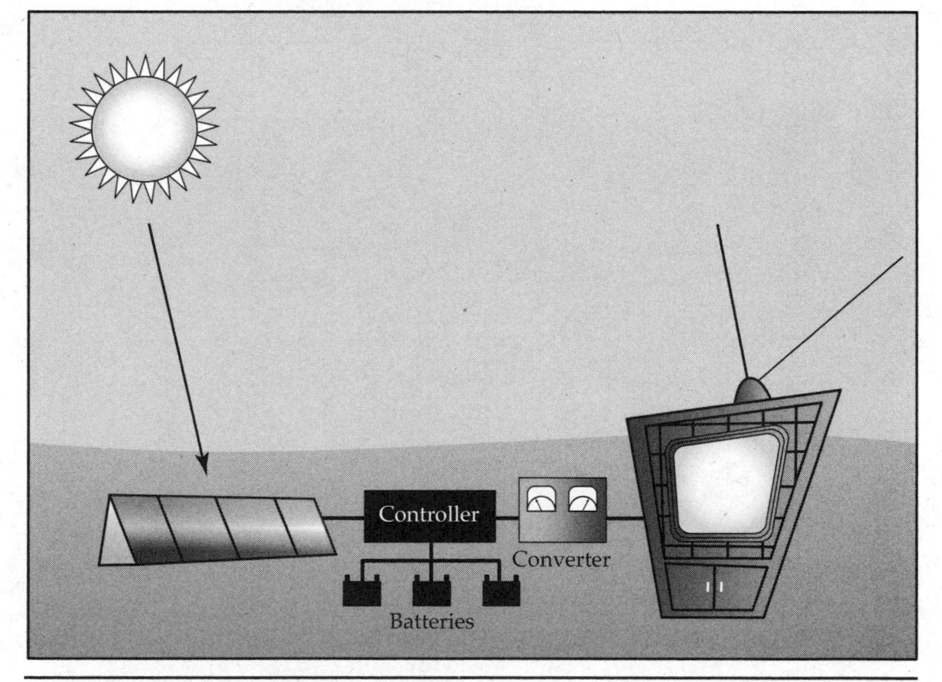

FIGURE 12-10 Independent solar system
http://www1.eere.energy.gov/tribalenergy/guide/images/off_grid_ac_solarsystem.gif

The electrical grid in the United States is predicted to require $2 trillion in upgrades by 2030 to maintain current levels of reliable energy (http://www.brattle.com/NewsEvents/NewsDetail.asp?RecordID=568). Individual home solar energy systems would eliminate this and future costs of maintenance, development, and new power stations. These types of independent systems are available for energy, water, and sewage, and all dramatically reduce pollution and carbon dioxide emissions.

Many technologies are available today and others are being developed every day to provide for energy now and in the future. Your job today is to take part in the solar revolution.

APPENDIX

Standards, Conversions, and Green Reference

This appendix reviews common terms and standards used in this book, plus references to more information.

Basic Units of Energy

Understanding electricity and how electricity is measured is imperative to your understanding of your solar energy system. Let's begin with the basic systems of electrical measurement.

A *volt* is the basic unit of electrical potential. One volt is the force required to send 1 ampere of electrical current through a resistance of 1 ohm.

A *watt* is a unit of power equal to 1 joule per second, the power dissipated by a current of 1 ampere flowing across a resistance of 1 ohm. Most people relate to the measurement of watts for light bulbs and small home appliances. Most compact fluorescent light bulbs, for example, are 4 to 40 watts, while average incandescent bulbs can be 25 to 150 watts, and a hair dryer is 1500 to 2500 watts.

Let's consider a 100-watt incandescent light bulb to determine how much energy it consumes. A *kilowatt* equals 1000 watts, so this light bulb supplies 0.10 of a kilowatt of power. A *kilowatt hour* equals the amount of power used by 1 kilowatt in one hour. So, if you turn on your 100-watt light bulb for 10 hours, you will have used 1 kilowatt total (0.10 × 10).

Heat Loss and Gain Measurements

Several terms are used in measuring the effectiveness of heat loss and gain from windows, doors, skylights, and insulation.

Btu's

The British thermal unit is basically the amount of heat energy made available by consuming a product. In a passive solar system, the amount of Btu's gained or lost determines the system's efficiency.

- 1 watt is approximately 3.413 Btu/h
- 1000 Btu/h is approximately 293 watts

Here are some examples of Btu's: Burning a gallon of gasoline creates approximately 125,000 Btu's of energy. The amount of energy required to melt one short ton of ice in 24 hours is 12,000 Btu/h. Converted to watts, this equals approximately 3.51 kilowatts.

A *therm* is a unit of energy used in the United States and the European Union: 1 therm equals 100,000 Btu's. One therm is approximately the energy equivalent of burning 100 cubic feet of natural gas.

U Factor

The U factor is the rate at which a window, door, or skylight conducts non-solar heat flow—that is, the rate at which it loses heat. The U factor is rated in Btu's and may refer to the glass, the glazing, or the entire unit. The lower the U factor, the more efficient the window or door. If you want to use the sun's heat energy, as in passive solar, you may want to vary the flow of energy into your home.

Solar heat gain coefficient or SHGC (Figure A-1) is the amount of heat energy transmitted through a window or door. The lower the ratings number, the more efficient the unit.

Light to Solar Gain

Light to solar gain (LSG) is a relative efficiency term that rates the amount light gained and the amount of heat blocked. The higher the ratings number, the better the performance.

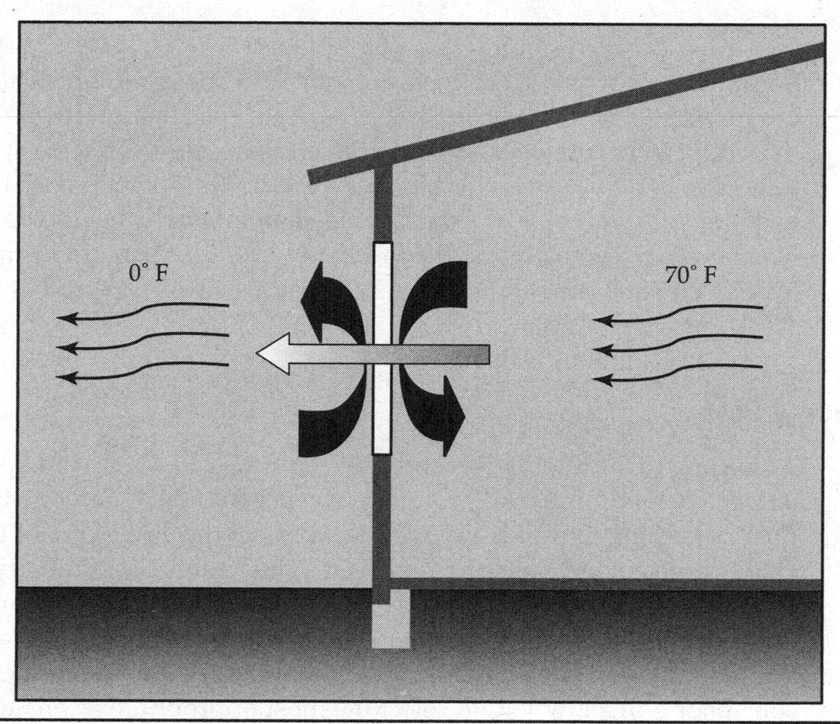

FIGURE A-1 How energy flows through a window
http://resourcecenter.pnl.gov/cocoon/morf/ResourceCenter/dbimages/full/10.jpg

R Value

The R value is the thermal resistance to heat: the higher the number, the better the resistance. This factor is often associated with insulation, but it also pertains to windows and skylights for passive solar projects.

Temperature and Humidity

Air temperature and water vapor in the form of humidity are also important to solar home systems. *Relative humidity* is the amount of water vapor that exists in the air at a known temperature. The amount of water contained by the air is relative to the temperature.

The *dew point* is related to relative humidity. The dew point is the point at which water will condense from the air at a constant barometric pressure. A relative humidity reading of 100 percent indicates that the dew point is equal to the current temperature. When the dew point falls below freezing, it is called the frost point.

Barometric Pressure

Barometric pressure is the measurement of atmospheric pressure—the force exerted against an object at a certain sea level. Barometric pressure is important to homeowners because the pressure differential (the pressure difference between the inside and the outside of the home) is one of the factors that causes air to flow from a home, therefore reducing the home's efficiency. Passive heating and cooling projects are dependent upon the relative climate, temperature, and barometric pressure in your area.

Energy Star

According to the Energy Star website: "ENERGY STAR is a joint program of the U.S. Environmental Protection Agency and the U.S. Department of Energy helping us all save money and protect the environment through energy efficient products and practices." The Energy Star rating system is based on statistical averages and assumptions calculated according to how much energy an appliance uses. The lower the Energy Star number, the more money you will save. Understanding your requirements as a consumer will allow you to select the most appropriate appliances. You can visit the Energy Star website and check products and ratings: http://www.energystar.gov/index.cfm.

In 1992 the EPA introduced Energy Star as a voluntary labeling program designed to identify and promote energy-efficient products to reduce greenhouse gas emissions. Computers and monitors were the first labeled products. Today, the Energy Star label appears on more than 50 product categories, including major appliances, office equipment, lighting, and home electronics. EPA has also extended the label to cover new homes and commercial and industrial buildings (http://www.energystar.gov/index.cfm?c=new_homes.hm_index).

By using Energy Star products, you can save hundreds of dollars' worth of energy use per year, with similar savings of greenhouse gas emissions, without sacrificing features, style, or comfort. Look for household products with the blue Energy Star label (Figure A-2). To qualify, these appliances must meet strict energy efficiency guidelines set by the EPA and U.S. Department of Energy (http://www.energystar.gov/index.cfm?c=products.pr_how_earn).

Use the Energy Star label to compare similar sized models. Then think about how you will use the appliance. Check any additional information or rating systems and choose the appliance that best fits your requirements. Energy Star labeling is excellent for most consumers, but knowing

FIGURE A-2 Energy Star label
www.energystar.gov

how you use appliances is another important consideration in choosing new appliances.

> Energy Star is creating new standards for energy efficiency and a more effective rating system. The new standards and developments can be found at http://www.energystar.gov/index.cfm?c=new_specs.new_prod_specs. The current standards are available at http://www.energystar.gov/index.cfm?c=prod_development.prod_development_index.

The Energy Star tax credit label (Figure A-3) identifies the fact that an Energy Star product is eligible for a tax credit or rebate.

FIGURE A-3 The Energy Star tax credit label
http://www.energystar.gov/index.cfm?c=windows_doors.pr_taxcredits

The EPA offers tools and resources to help you plan and undertake projects to reduce your energy bills and improve home comfort. See http://www.epa.gov/greenhomes/ReduceEnergy.htm for more information.

EnergyGuide

The EnergyGuide label (Figure A-4) assists consumers in choosing the most energy-efficient product for their homes.

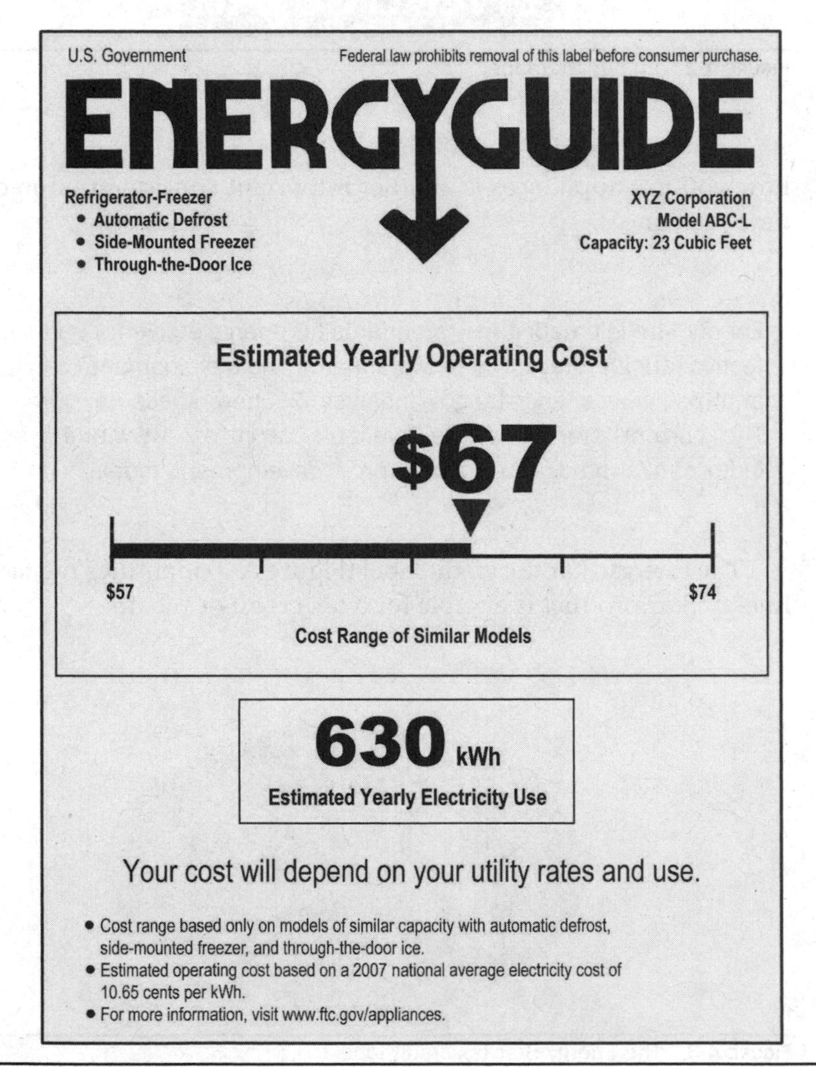

FIGURE A-4 EnergyGuide label
http://www.ftc.gov/opa/2007/08/elabel.jpg

The EnergyGuide label provides yearly operating costs and electric use information about the specific product to which it is attached. These products are subject to minimum efficiency standards set by the federal government. Consumers will find these labels only on appliances that meet the Energy Star requirements.

Some products, such as televisions, clothes dryers, ranges and ovens, and space heaters already have to meet federal minimum efficiency standards and are exempted from the EnergyGuide program. The amount of energy these products use does not vary substantially from model to model.

Other Ratings Systems

Energy Efficiency Rating (EER) is the Air-Conditioning and Refrigeration Institute standardized rating system. This rating measures steady-state efficiency: the higher the EER number, the more effective the product.

The Seasonal Energy Efficiency Rating (SEER) is available for residential central air conditioners and is generally considered a more reliable indicator of the overall energy efficiency of the unit than the EER. The higher the SEER number, the more effective the product.

Tax Credits and Rebates

In addition to products, services may also be entitled to tax credits or rebates. Products that require specialty installation may be available for tax credits:

- Water heaters
- Central air conditioning
- Heat pumps
- Solar panels
- Wind energy systems
- Heating systems
- Fuel cells
- Wood or pellet stoves

Other Sources of Information

- **Federal Tax Credits for Consumer Energy Efficiency:** http://www.energystar.gov/index.cfm?c=tax_credits.tx_index
- **Energy Terminology:** http://www.consumerenergycenter.org/glossary/
- **U.S. Department of Energy:** http://www.energy.gov/
- **U.S. Environmental Protection Agency:** www.EPA.gov
- **EnergyGuide:** http://www1.eere.energy.gov/consumer/tips/energyguide.html
- **Energy Star Program:** http://www.energystar.gov/
- **Solar Energy Industries Association:** http://www.seia.org/
- **Database of State Incentives for Renewable Energy (DSIRE):** www.dsireusa.org
- **National Association of Regulatory and Utility Commissioners (NARUC):** www.naruc.org
- **Solar Energy Technologies Program:** www.eere.energy.gov/solar
- **National Renewable Energy Laboratory:** www.nrel.gov/ncpv
- **Million Solar Roofs:** www.millionsolarroofs.com
- ***Consumer's Guide to Buying a Solar Electric System*:** www.nrel.gov/docs/fy04osti/35297.pdf

I have found all of these websites to be credible and up to date. With technology changing so rapidly, it is as important that you have up-to-date information as well as reliable and credible information. Every attempt has been made to validate all of the websites and resources referred to in this book. If you have any additional links or reference information that you would like to see in future printings, please contact me at Solarinfo@exploresynergy.org.

Index